IN THE FOOTSTEPS OF ZARAFA, FIRST GIRAFFE IN FRANCE

IN THE FOOTSTEPS OF ZARAFA, FIRST GIRAFFE IN FRANCE

A Chronicle of Giraffomania, 1826–1845

Olivier Lebleu

Translated by Cynthia T. Hahn

ROWMAN & LITTLEFIELD
Lanham • Boulder • New York • London

Published by Rowman & Littlefield
An imprint of The Rowman & Littlefield Publishing Group, Inc.
4501 Forbes Boulevard, Suite 200, Lanham, Maryland 20706
www.rowman.com

6 Tinworth Street, London SE11 5AL, United Kingdom

Copyright © 2020 by The Rowman & Littlefield Publishing Group, Inc.

All rights reserved. No part of this book may be reproduced in any form or by any electronic or mechanical means, including information storage and retrieval systems, without written permission from the publisher, except by a reviewer who may quote passages in a review.

British Library Cataloguing in Publication Information Available

Library of Congress Cataloging-in-Publication Data

Names: Lebleu, Olivier, 1966- author. | Hahn, Cynthia T., 1961- translator.
Title: In the footsteps of Zarafa, first giraffe in France : a chronicle of giraffomania, 1826-1845 / Olivier Lebleu ; translated by Cynthia T. Hahn.
Other titles: Avatars de Zarafa. English
Description: Lanham : Rowman & Littlefield, [2020] | Includes bibliographical references and index.
Identifiers: LCCN 2020008464 (print) | LCCN 2020008465 (ebook) | ISBN 9781538142233 (cloth) | ISBN 9781538142240 (paperback) | ISBN 9781538142257 (epub)
Subjects: LCSH: Giraffe—France—History—19th century. | Zoo animals—France—History—19th century.
Classification: LCC QL737.U56 L43513 2020 (print) | LCC QL737.U56 (ebook) | DDC 599.6380944—dc23
LC record available at https://lccn.loc.gov/2020008464
LC ebook record available at https://lccn.loc.gov/2020008465

CONTENTS

AUTHOR'S NOTE		vii
FOREWORD BY BRUCE D. PATTERSON		ix
PREFACE		xi
ACKNOWLEDGMENTS		xiii
1	THE GIRAFFE BEFORE ZARAFA: A Polymorphous Monster	1
2	ILLUSTRIOUS GODPARENTS: A Predestined Animal	13
3	THE PASHA'S GIFT: Diplomacy at Stake	25
4	ZARAFA IN FRANCE: A Royal Triumph	37
5	THE ANIMAL UNDER A MAGNIFYING GLASS: An Aberration of Nature	53
6	BETWEEN SCIENCE AND RELIGION: An Animal at a Price	67
7	TRADE AND GIRAFFOMANIA: Marketing Materials	93
8	POLITICAL CARICATURE: A Satirical Weapon	109
9	ZARAFA'S LEGACY: An Existential Fable	141

CONTENTS

NOTES 157
BIBLIOGRAPHY 163
INDEX 173
ABOUT THE AUTHOR AND TRANSLATOR 185

AUTHOR'S NOTE

Since "style is personal," the correspondence cited in this work is often in extenso, in order to free the reader to follow the thought of its author, whether luminous or tortuous, respecting its original distinctiveness as much as possible in the translation.

FOREWORD

The giraffe occupies a special place in our hearts and imaginations. Traditionally, giraffes have been considered a single living African species, but insights from DNA may indicate that they represent as many as eight or nine distinct evolutionary lineages. Moreover, the fossil record documents seven additional prehistoric species ranging as far afield as India and Japan. Giraffes' long neck and long legs make them far and away the most gracile of the megaherbivores (plant-eating mammals that weigh more than a metric ton), and they tower over even the largest pachyderms. Whereas elephants, hippos, and rhinos all display menacing tusks and horns, giraffes have inoffensive ossicones, skin-and-hair-covered knobs found on the heads of both male and female giraffes. While most other giant mammals are rather indiscriminate feeders, scarfing up practically anything, giraffes impress observers with their careful selection of leaves, gathered with a remarkably long and prehensile tongue. And because leaves are very hard to digest and require hordes of microbes to release their nutrients, passage through the giraffe's four stomachs and 70 meters (230 feet!) of digestive tract takes nearly two days and efficiently (and utterly) transforms a bushel of leaves into a large handful of hard pellets.

Giraffa display marvelous adaptations to their savanna ecosystems and have achieved considerable evolutionary success over the last eleven million years, but they face unprecedented challenges in the twenty-first century. Although giraffes are still widespread in eastern and Southern Africa, they have apparently gone extinct in seven range states (Burkina Faso, Eritrea, Guinea, Mali, Mauritania, Nigeria, and Senegal) and become rare in others. Over the past thirty years, the number of mature individuals has declined by 36 to 40

FOREWORD

percent, and the IUCN (International Union for Conservation of Nature) estimates that fewer than seventy thousand adults remain.[1] Shrinking populations evident in this species-wide assessment are even more worrying when the various lineages are considered separately: populations of the Kordofan giraffe (in the central Sahel) have declined by 46 percent, the reticulated giraffe (in Kenya, Ethiopia, and Somalia) by 80 percent, and the Nubian giraffe (found in Ethiopia and South Sudan) by 97 percent! The threats to giraffes vary in severity and importance by region, but all stem from human activities: (1) habitat loss and conversion, (2) civil unrest, (3) illegal hunting (poaching), and (4) changing ecological conditions. So resolving these challenges and reversing these declines clearly depend on humans.

In 1968 a young Senegalese forestry engineer named Baba Dioum spoke at a meeting of the IUCN, and observed, "In the end we will conserve only what we love, we will love only what we understand, and we will understand only what we are taught." His summation offers a concise but compelling argument for the importance of this collaboration of Olivier Lebleu and his translator, Cynthia Hahn. In this book, they document the excitement and wonder that stemmed from seeing the first live giraffe in modern France and the knowledge and understanding it engendered. Lebleu and Hahn offer us a very timely perspective—would that it now trigger the type of wonder and emotion that Zarafa inspired nearly two centuries ago, as giraffes need our attention and support now more than ever.

Bruce D. Patterson, PhD
Field Museum of Natural History
Chicago, Illinois

PREFACE

It was Gabriel Dardaud, then director of the Agence France-Presse for the Middle East, living in Cairo, who accidentally unearthed the extraordinary story of the first giraffe in France, starting with the creature's entry into Marseille on October 23, 1826. Only a few archivists, museum curators, or private collectors had preserved evidence of the passage of this gentle-eyed giant. Just after World War II, Dardaud, a former war correspondent surely found a welcome escape from current events through this compelling discovery.

Dardaud was passionate about Zarafa, as evidenced in his compilation of sources, collection of relics, and talks given; thirty years later, for the occasion of an exhibit at the museum in Sceaux, he put together a reference work that remains a jewel with regard to its scholarship and humor. In 1999 the American Michael Allin recalled to French memory the journey of this giraffe, from the Kordofan desert right up to the establishment of this collection at the Muséum national d'Histoire naturelle in Paris.

So why revive the memory of this beautiful foreigner, whose preserved hide is now the pride of the Muséum d'Histoire naturelle in La Rochelle? Because no book had yet rendered a full account of the effects of the veritable giraffomania that Zarafa unleashed in the second quarter of the nineteenth century. The French people latched onto this gift destined for a king from a pasha. Promenaded triumphantly from Marseille to Paris as a marketable, living icon, the giraffe then saw its image proliferated across all sorts of items, in two or three dimensions.

Due to her extravagant contours and amiable character, Zarafa became the subject of much fantasy. Everyone projected their strongest preoccupations upon this never-before-seen animal. Like her Egyptian caretakers, Zarafa was led

PREFACE

to encounter scientists, religious, merchants, politicians, artists.... The result was an extraordinary bounty of literature and iconography that the present work aims to present. Images from this time may be layered one upon another, artistic and commercial products are displayed, and exchanges of correspondence are laid out according to a chronology that describes the arc of this "giraffic fever." Our ambition was to set down on paper, for posterity's sake, the museum that has yet to be built in homage to this Egyptian orphan who was to become queen of France, in spite of herself, for some twenty years.

The original, French edition of this work was published by Arléa in 2006. Six years later, a Franco-Belgian coproduction, *Zarafa* (2012), appeared on the big screen, a splendid, joyfully mischievous, animated film by Rémi Bezançon and Jean-Christophe Lie that sidestepped facts from the historical account. This animation for children was preselected for the Academy Awards, and the giraffe narrowly missed being brought to Hollywood! In 2017 I wrote, published, and acted in a six-person theatrical comedy titled *Le Talisman de la Girafe*, composed entirely in alexandrine verse. And I dream of a musical comedy for the big stage! Will Zarafa have her day in America? Will she come to represent her species as a whole on the global stage and become a standard-bearer?

More than anything, I wish for giraffes—considered to be in danger of extinction since 2016—to continue living upon this planet earth. If they were to disappear, humanity would lose more than the most beautiful representatives of African fauna: it would lose part of its very soul.

Olivier Lebleu
November 2019

ACKNOWLEDGMENTS

Archives départementales des Bouches-du-Rhône
Archives municipales de Marseille
Association Vieux Papier:
    Mr. Thierry Depaulis, president
Bibliothèque de l'école vétérinaire de Lyon:
    Ms. Marie Viau, Vétagro Sup
Château d'Arlay:
    Mr. Alain de la Guiche
Editions Arléa:
    Ms. Lucie Lesvenan, executive assistant
Groupe Pierre Frey
Mairie d'Auxerre
Mairie de Beaune:
    Ms. Anne Caillaud, assistant, Department of Culture and the Promotion of Cultural Heritage
    Ms. Delphine Cornuché, Museums Department
Mairie d'Étampes
Musée barrois de Bar-le-Duc:
    Ms. Claire Paillé, curator
Musée Carnavalet:
    Mr. David Simonneau, curator assistant, Graphic Arts Department
Musées de Beaune:
    Ms. Marion Leuba, curator
Musée de Bouxwiller
Musée de Fécamp

ACKNOWLEDGMENTS

Musée de la Toile de Jouy
Musée départemental d'art ancien et contemporain d'Épinal
Musée des arts et traditions populaires
Musée du Domaine départemental de Sceaux:
    Ms. Marie-Noëlle Mathieu, assistant curator
    Mr. Antoine Bourroux, librarian, Center for Documentation and Photography
Muséum d'Histoire naturelle de la Rochelle:
    Ms. Élise Patole-Édoumba, curator
Musée intercommunal d'Étampes:
    Mr. Sylvain Duchêne, museum manager
    Mr. Thomas Crosnier, collections curator
Muséum national d'Histoire naturelle
Photothèque des musées de la ville de Paris

PERSONAL ACKNOWLEDGMENTS

Jean-Marie Aynaud (for the use of his father's collection)
Ms. Marie-Laure Daudet
Ms. Sophie Kalbach
Ms. Adriana Rosado-Bonewitz
Mr. Rand Smith

1

THE GIRAFFE BEFORE ZARAFA

A Polymorphous Monster

The most ancient representations of the giraffe date back to prehistory. In 1909 an officer of the Department of North African Indigenous Affairs, one Captain Cortier, was the first to note the existence of caverns in the Sahara Desert with polychromatic etchings depicting stylized images of giraffes. In 1928 Conrad Kilian, a Frenchman of Alsatian origin who discovered oil in the Sahara, found frescoes in Tassili representing a giraffe hunt. The ethnologist Henri Lhote in 1956 undertook the systematic, scientific study of archeological remains in the Sahara, cataloguing more than 150 sites; some of the sites preserved frescoes dating back to 6000 BCE. A painted engraving on a cliff of the In Djaren wadi shows a warrior holding a kind of javelin in his right hand and in the other, a giraffe on a leash. The link between giraffes and people is older than we might think.

The mystery of the giraffe begins with its name, or rather, its names. Well before the French language had adopted the term *girafe*–often written *giraffe* in Zarafa's time–derived from the Arabic *Zerafa*, meaning "charming" or "pleasant," the animal had inherited a variety of patronyms. These multiple identities stemmed as much from the cultural diversity of its observers as from their diverse perceptions of this unique animal.

Moses identified it with the name *Zemer* in Deuteronomy, declaring it theologically edible: "Do not eat any abomination. These are the mammals that you may eat: the ox, the sheep, the goat, the gazelle, the deer, the antelope, the ibex, the chamois, the bison, and the giraffe. You may thus eat every animal that has a true hoof that is cloven into two parts, and which brings up its cud."[1] In the past, Ethiopians called it *nabis*, *nabu*, or *nabuna*, from the root *naba*, which meant

CHAPTER 1

"to be high up." For Egyptians, the hieroglyph representing a giraffe signified not only "to see" but also "to see far" or "to foresee."

The Greeks, reflecting that it brought together the traits of a horse, a panther, a leopard, and a camel, called it *pardion, hippardion* ("horse-panther"), or *camelopardalis* ("camel-panther"). Aristotle only briefly mentioned it, while Agatharchides, a geographer living around the year 104 CE, gave a first description of the animal: "With the Troglodytes lives also the animal the Greeks named camelopardalis, a compound noun expressing the dual nature of this quadruped. It has the markings of a leopard, the girth of a camel, and it is extraordinarily tall. Its neck is long enough for it to graze from treetops."[2]

The poet Oppian, born around the year 180 CE, sings of the giraffe in his *Cynegetica*, with the accuracy of a naturalist:

> Now I pray for you to tell me, oh my Muse
> Of these double animals of a nature confused:
> The camel-panther. Oh, how many different varieties
> Of beasts of the air, and fishes of the sea,
> Father of the universe, Jupiter, great and wise,
> You have invented for us, and all for our use!
> You have even sown a strange offshoot of camel,
> From which camel-panther has sprung, gentle animal,
> Of agreeable beauty, bald on top of its head,
> Its body patterned, feet long and neck grand;
> With its short ears and large heels,
> Of unequal measure legs and feet,
> Smaller behind and as if lame, slowly
> Struggles to circumvent any obstacle. Two feeble horns
> Spring forth from its head, like garden forks;
> Its mouth wide as a deer, where its teeth, fine and
> White as milk, are firmly gathered;
> As for the rest, an ardent eye and short tail at back,
> That of a doe or buck whose tail is tipped black.[3]

In the fourth century the writer Heliodorus of Emesa mentions a giraffe in his novel, *Aethiopica, or Theagenes and Chariclea*; in addition, he was the first to note its peculiar gait: "Its walk is also strange and contrary to every other kind of animal, on earth or in the sea; it doesn't stir one foot from one side and then another from the other side, but advances its two right feet together and the two left also together. Thus, it is always walking with one side suspended in the air."[4]

Latin adopted the Greek terminology of the giraffe, or referred to it as *ovis fera*, "wild sheep," due to its passive demeanor. The historian Pliny the Elder recounted the odd animal's arrival in Rome:

> ... with a neck the size of a horse, the feet and legs of a bull, the head of a camel and white spots scattered across a tawny background, which has given it the name of *camelopardalis*. The first giraffe was seen in Rome during circus games held by the dictator Caesar (Year of Rome 708); since then one has been seen from time to time. This animal is more remarkable due to its extraordinary appearance than by its wild nature; it has thus received the name of wild sheep.[5]

Under the reign of Julius Caesar in 46 BCE, giraffes were regularly exhibited in Rome, unleashing the irony of Horace with regard to his contemporaries: "How Democritus would laugh if he were still upon the earth, with a wide-mouthed, gaping multitude, gazing upon either a white elephant, or a monster, half camel, half panther!"[6] While the games celebrated the first thousand years of the foundation of Rome, Emperor Philip the Arab, successor of Gordian III, offered the people of Rome ten giraffes at a time. Like the other wild animals, these were surely massacred in the circus. The chronicler Eusebius Pamphili noted that these barbaric festivities lasted three days and three nights.

In the tenth century, the Arab geographer Ibn al-Faqih evoked an animal anthology with regard to the giraffe: "The size of a camel, the head of a deer, the skin of a panther, the hooves of a cow, and the tail of a bird." Scholar Zakariya al-Qazwini included curious details in his *Wonders of Creation* in the thirteenth century: "Feet like those of a nine-year-old camel and with the tail of a gazelle." His zoological treatise resembles a tale from *The One Thousand and One Nights*. According to him, "a male hyena must mate with an Abyssinian female camel" and, "if the offspring is a male, it must then mate with a wild cow, which then gives birth to a giraffe."

Most of the European naturalists of the Middle Ages and from the Renaissance give the giraffe an approximate description, contenting themselves with repeating others' accounts without being able or wishing to verify the source. The best descriptions are eyewitness accounts. Marco Polo, stopping in Zanzibar in 1295, recounted that its inhabitants had a "kind of animal they call Giraffa. It has a collar as long as three paces; it has much longer front legs than hind legs; it has a small head and its body sports several colors, such as white, red and patched: this animal is gentle and harmless."[7]

The trend began to develop of giving a real giraffe or some representation of it as a gift. During the thirteenth century, King Saint Louis received a crystal miniature of a "beast called Orafle," while Frederick II of the Hohenstaufen dynasty in

CHAPTER 1

Palermo received a flesh-and-blood giraffe described obliquely by the theologian monk and philosopher Albertus Magnus under the name of *oraflus* or *anabula*.

In ancient China the giraffe was called *kilin*, described as a fabulous animal, with the body of a buck, the tail of a cow, covered in scales, and sporting a single horn, emblem of supreme virtue. In 1414 Zhu Di, third emperor of the Ming dynasty, jealous of the king of Bengal, Saifuddin, who had received a giraffe for his inauguration, demanded one for his imperial garden. The mission was accomplished three years later, via some ambassadors specially dispatched to Africa. On November 28, 1488, in Lyon, France, a religious brother by the name of Nicole Le Huen published a representation of the giraffe, among other legendary animals, in a work entitled, *Des sainctes pérégrinations de Jhérusalem et des anvirons et des lieux prochains. Du mont du Synay et de la glorieuse Katherine: Cet ouvraige et petit livre contenant du tout la description ainsi que Dieu a voulu le donner à congnoistre, par frère Nicole le Huen, humble professeur en sainte théologie, religieux à la mère de Dieu nostre Dame des carmes du couvent de Ponteaux de Mer et de la feu reine Charlote que Dieu absolve, confesseur et dévot chapelain et le vrement perpétuel subgect et orateur*. This is probably the first French illustration of the animal (figure 1.1).

During the Renaissance, writers took up the Latin denomination, designating the giraffe under the name of camelopard, cameleopard, or camel-leopard. Leo Africanus (al-Hasan ibn Muhammad al-Wazzan al-Fasi), an Arab geographer, depicted it in his *Description of Africa*:

> This animal is so strange and so wild before capture that it is rarely seen, because it hides well in the woods and desert lands where it dwells, where other animals never go, and as soon as it perceives a man, it tries to maintain its distance; but it is caught easily as it is slow to run.... It is the gentlest of all beasts to govern... and oftentimes I guided it, without ever seeing it try to bite or kick me.... Those who hunt them choose only to capture those that are still very small, from places where they have just been born.[8]

In 1551, in his *History of Animals*, the doctor and naturalist Conrad Gesner was content to record the description given by his predecessors and to copy the drawing by Brother Le Huen.[9] His contemporaries, however, included eyewitnesses. In 1588 the naturalist Pierre Belon also described it more fully. The detail he gave of the front legs could not have been invented:

> It is a very beautiful animal of the sweetest nature, similar to a ewe and as pleasant as no other wild animal.... It could not graze upon the earth while standing, without spreading its front legs far apart. And even then, with great difficulty. This is why it is

Figure 1.1. First French representations of the giraffe; drawn by Nicole Le Huen, 1488.
In Joly, N., *Notice sur l'histoire, les mœurs et l'organisation de la girafe*, Toulouse, 1845.

Figure 1.2. A monster, according to Ambroise Paré, in 1585.
In Paré, Ambroise, *Les Œuvres d'Ambroise Paré, . . . 8e édition, revues et corrigées en plusieurs endroicts et augmentées d'un fort ample traicté des fiebvres . . . nouvellement trouvé dans les manuscrits de l'autheur*, G. Buon, Paris, 1585. Bibliothèque de l'Ecole Nationale Vétérinaire de Lyon, Public Domain.

easy to believe that it does not live in fields except from the branches of trees, having such a long neck, so long that its head could be as high as a half spear.[10]

Such anthropomorphic descriptions would resume during Zarafa's time.

When it was not the giraffe moving in the direction of human civilization, it was the explorer who sought out the giraffe and recounted ever more detailed stories about the animal. The Spanish traveler Luis del Marmol was fascinated by its laid-back demeanor: "It walks solemnly without reacting or flinching with regard to anything. Africans say that it has evolved from several species. It avoids other animals in the woods and flees from humans. They are taken small from the places the mothers frequent."[11] We note this remark, as it remains true for the future Zarafa: to favor her chances of domestication, the giraffe was captured under its mother, not yet weaned.

In 1624 the giraffe found its place among the heavens as a foreshadowing of its future stardom. Jacob Bartsch, first assistant to Johannes Kepler's son-in-law at the time, named the Giraffe constellation in order to fill in an undescribed area between the most ancient constellations named by the Greeks. About 757 square degrees in size, located within the Northern Hemisphere between the Big Dipper and Cassiopeia, it is rather difficult to see due to the feeble magnitude of the stars it comprises.

In 1657 the drawings of John Jonston, doctor and naturalist of Scottish origin, appeared completely fantastical.[12]

In 1662 Michel Baudier recounted in his *General History of the Harem* that he saw a giraffe in Constantinople during festivities celebrating the circumcision of Mohammed III, and he made this noteworthy but exaggerated comparison: "The front legs are four or five times higher than those of the hind legs, so much so that its natural posture is akin to a goat pitched upward in a tree, grazing upon the offshoots."[13]

In Richelet's *Dictionnaire*, whose first edition dates from 1688, the article on the "Giraffe" summarizes the knowledge of the time while adding a few homegrown remedies: "It is a kind of camel that stems from the leopard. Its neck is up to seven feet in length, with tufts of coarse hair, akin to a horse's mane. This animal is gentle and trainable, although wild. Its horns and nails may be used to treat epilepsy and diarrhea, if grated, crushed and then ingested."

But the term *giraffe* also designated "a kind of gray cloth mixed with a bit of white, apt to produce a good fur, because it is made of the skin of the animal called giraffe."[14]

In 1768 the giraffe of the encyclopedists was portrayed rather close to reality, aside from the horns[15] (figures 1.3a and 1.3b). In 1776 and 1787 Buffon himself

Figure 1.3. These visions of Dutchman Vosmaer from 1747 are among the most realistic.

In Jonston, Jan, *Historiae naturalis de quadrupedibus libri, cum aeneis figuris Johannes Jonstonus, medicinae doctor concinnavit*, Amstelodami, Apud Joannem Jacobi Fil. Schipper, 1747. Bibliothèque de l'Ecole Nationale Vétérinaire de Lyon, Public Domain.

CHAPTER 1

had not studied the giraffe alive, thus his description remains incomplete, erroneous, and philosophically debatable:

> The Giraffe is among the top most beautiful and tallest animals, which, without being dangerous, is also one of the least useful. The enormous disproportion of its legs, of which the front are twice as long as the hind ones, renders its movement difficult; its body has no seat; its walk is halting, its movements slow and constrained; it can neither flee its enemies when free, nor serve its masters when domesticated: the species is not prolific and has always been confined to the deserts of Ethiopia and several other provinces of Southern Africa and Southeast Asia.[16]

In reality the giraffe is robust and hardy, and its legs are not that enormous. It can break a lion's skull with its kicks; the horns are but the extension of its frontal bones, etc. The drawing he presented disproportionately displays a gigantic giraffe, or an autochthonous Lilliputian (figure 1.4).

Contrary to rumor, stemming from ignorance or nationalistic enthusiasm, Zarafa would not be the first giraffe to grace European soil. Three living giraffes had already made the trip. The first, as noted, was sent in the thirteenth century to Frederick II, king of Sicily and Germanic emperor, from the sultan of Egypt in return for the gift of a white bear. The second was offered by Sultan Biba to Manfred, natural son of Frederick II. The third was given in 1486 by the Mamluk Qaitbey to Lorenzo de' Medici, who put it in the zoo in Florence and whose image was reproduced by Giorgio Vasari on frescoes decorating the Palazzo Vecchio. The Italian giraffe was envied by those on the other side of the Pyrenees; in a letter dated April 15, 1489, Anne de Beaujeu, daughter of French king Louis XI, used her charming pen to solicit a domestic visit by the prince of Florence: "I beg of you to send me the animal giraffe which is the animal in all the world I have the greatest desire to see and if there is something that I might be able to do for you in return I will strive to do it with all my heart."[17]

It is easy to understand Lorenzo's refusal; he must have thought that extending the law of courtesy was not worth the risk of losing the precious animal.

Unable to count on the generosity of a foreign monarch, explorers became hunters. It was easier to bring back a skin and a sack of bones than a living giant. The animal gave French explorer François Levaillant a hard time: "We saw the Giraffe traverse the plain; it moved at a light trot, not appearing to be in a hurry. We galloped after it, and from time to time fired upon it with our guns, but gradually it gained such distance on us that after having pursued it for three hours, and forced to stop because our horses were out of breath, we lost sight of it" (figure 1.5).

Figure 1.4. Buffon's rendering of a spotted version, 1776.

In Buffon, Georges-Louis, *Histoire naturelle générale et particulière servant de suite à l'histoire des animaux quadrupèdes par M. le Comte de Buffon, intendant du Jardin et du Cabinet du Roi, de l'Académie française, de celle des sciences*, etc., Impr. Royale, Paris, 1776–1787. © Muséum d'Histoire naturelle de La Rochelle / Lezard Graphique.

Figure 1.5. Female giraffe, by Levaillant.

In Levaillant, François, *Premier voyage dans l'intérieur de l'Afrique par le cap de Bonne-Espérance*, Paris, 1790; *Second voyage dans l'intérieur de l'Afrique*, Paris, 1796. © Muséum d'Histoire naturelle de La Rochelle / Lézard Graphique

2

ILLUSTRIOUS GODPARENTS

A Predestined Animal

The French Zarafa was a gift from Egypt to France. At the origin of its legendary destiny, three exceptional men came together in the land of the former pharaohs: the Kurd Muhammad Ali (1769–1849), the Frenchman Étienne Geoffroy Saint-Hilaire (1772–1844), and the Italian Bernardino Drovetti (1776–1852).

AN ENLIGHTENED TYRANT: MUHAMMAD ALI

It was 1798. Napoléon Bonaparte had just disembarked in Egypt. At first well-received, he soon came into conflict with the Mamluks, a dynasty of former slave-mercenaries who governed the country but without reigning over it. They were only vassals of the sultan, supreme leader of the Ottoman Empire established in Constantinople. As the sultan knew that the Mamluks were facing a tough battle against French invaders, he decided to send an army to help them. Within its ranks was a young general of Kurdish origin called Muhammad Ali. Thirty years of age, intelligent and ambitious, he was impressed by the organization and efficiency of the French army (figure 2.1).

After the French departure in 1800, the sultan was unable to reestablish order, and Muhammad Ali, assisted by prominent local individuals and henchmen, took advantage of this situation to defeat the Mamluks and take power. From the sultan, he obtained the title of pasha (governor) of Egypt in 1805, inheriting a country in a lamentable state of affairs. Since the sixteenth century, under the domination of Turkish pashas and then under the governance of Mamluk beys,

Figure 2.1. Muhammad Ali, an enlightened tyrant.
From the Memory of Modern Egypt Digital Archive. Bibliotheca Alexandrina, Public Domain, Egypt and United States.

Figure 2.2. The giraffe in a European depiction of the desert.
© Aynaud Collection.

the country had been progressively weakened and impoverished. Its admirable irrigation system had degraded due to lack of support, ruining agriculture and trade. Weakened by internal struggles within the Mamluk dynasty, the political institutions faced a severe decline.

The French invasion marked a turning point in the history of the country, and paradoxically brought Egypt into modernity, through a metamorphosis generated by Muhammad Ali. He had decided to create a modern state and army and to disengage Egypt from the Ottoman Empire. For this he would rely on French instructors and technicians, all the while integrating Egyptians into his army. Unscrupulous, he began to kill off the Mamluk leaders in 1811 and to exile the prominent men and Muslim religious leaders who had helped him attain power. With this new freedom, he then launched two large projects: to modernize the country and acquire new territory.

To finance his new army, Ali developed agriculture for export, namely, cotton and wheat. With oversight from French experts he had solicited, he built a network of irrigation and drainage canals and then constructed a dam to retain flood waters at the entrance to the Nile Delta, which assured Egyptians of water year-round as well as several harvests during the year. To give Egypt a strong place in the realm of international trade, he had the idea of a canal linking the Red Sea to the Mediterranean and of a railway between Cairo and the great port of Suez—projects that his successors would complete.

In tandem with the internal consolidation of his empire, Muhammad Ali sought to extend its borders. From 1811 to 1818, his troops fought the Wahhabites of Arabia and occupied the holy cities of Mecca and Medina, which conferred great prestige upon the pasha from around the Arab-Muslim world. Beginning in 1820, he sent his unruly soldiers to conquer Sudan, at the time called Kordofan, land of the future Zarafa. He built the capital, Khartoum, and planned to control the upper Nile and the succession of caravans from Central Africa, open new markets for his budding industries, and equip his army with black slaves. But in 1821, during the conflict between the Turks and Greek patriots, the visionary tyrant saw a dangerous drop in his international popularity.

A VERSATILE SCIENTIST:
ÉTIENNE GEOFFROY SAINT-HILAIRE

Étienne Geoffroy Saint-Hilaire was born in Étampes on April 15, 1772, "of an honorable family," but of little wealth, according to his son and biographer.[1] A

CHAPTER 2

young man with a delicate constitution, he cultivated a vigorous imagination. He first entered the seminary, which he left after a short time to pursue his passion in the natural sciences. His father allowed him to "come up" to Paris at age eighteen to study medicine, on the condition that he also study law. In less than a year, he obtained a degree in law, though he never practiced it. Biology was the first love of this multidisciplinary genius: at twenty years of age, he already displayed the courage that would always accompany his adventures as a man of science.

For example, in 1792 he contributed to the liberation of his former professor and friend, Abbott Haüy, imprisoned as a defiant priest. He convinced the head of the *Muséum national d'Histoire naturelle* (National Museum of Natural History), Daubenton, to alert the *Académie des Sciences*. Haüy was then released as "useful to the interests of science." However, almost all of his former instructors from the *Collège de Navarre* remained incarcerated in Saint-Firmin prison. Borrowing papers from a prison commissioner, Geoffroy gained access to the prison and begged the ecclesiastics to follow him out. But these men refused to save their lives, which would have created a greater risk for those detained with them. During the massacres that began at that time, this young man witnessed an elderly man being thrown from a window. The next night, Geoffroy mounted the prison wall in an agreed-upon place that was set the prior day; after waiting eight hours in this perilous spot, he helped twelve priests to escape. When daylight broke, the escape was still in progress, but this intrepid man would not leave his ladder until he had escaped from the gunfire. Several days later, executions ensued for those whom he could not save. Very affected by this, to the point of physical illness, Geoffroy returned to Étampes. After several months of convalescence, he returned to Paris, where Haüy strongly recommended him to Daubenton: "Love, help, adopt my young liberator."

Through his act of bravery, Geoffroy gained great notoriety in scientific circles and obtained a post as demonstrator with the Faculty of Natural History, replacing the resigning Lacépède. At age twenty-one, he was received among the twelve founders of the National Museum of Natural History and became president of the Department of Quadrupeds, Cetaceans, Birds, Reptiles and Fish. As a zoology professor, he was responsible for teaching a subject he knew little about at the time. And if that were not enough already, in executing an order of the Convention,[2] he founded the museum's zoo and became its director in year 2 (1793).

He was twenty-six when he and his friend and colleague Cuvier were proposed for Bonaparte's Egyptian campaign. Only Geoffroy accepted. In April

Figure 2.3. The giraffe walks and runs by ambling, raising both legs of the same side at the same time.
© Aynaud Collection.

1798 he left Paris in the company of his brother, Marc-Antoine, engineering captain, along with the naturalist Savigny and the painter Redouté.

With Egypt defeated, Geoffroy undertook a vast inventory of Egyptian fauna and contributed to founding the Institute of Liberal Arts and Sciences in Cairo. After Bonaparte's departure in 1799, he gave in to a moment of discouragement and confided in his letters that he had nothing left to accomplish on Egyptian soil. General Kléber, to whom the emperor had ceded power as he departed, was stabbed by a Muslim man. The young naturalist nevertheless took up his research once again. After the official French capitulation, the Oriental Expeditionary Force departed with English authorization in July 1801. These scholars then discovered to their great dismay that they had to render the fruits of their research to the new invaders. Without hesitation, Geoffroy declared to them that he preferred to destroy his collections rather than to be separated from them. "You seek celebrity. Well then! You can be sure that history will remember you, because you too will have burned down an Alexandrian library," he cried out to the astonished Englishman Hamilton.

CHAPTER 2

Figure 2.4. Hunting the giraffe, as imagined by the English.
© Aynaud Collection.

It was only in 1802 that Geoffroy returned to France, after surviving many perils. He brought with him the precious collections gathered in Egypt, and upon returning to the museum, he began describing them. By age thirty-five, Geoffroy had not only contributed to enriching the sciences through his descriptions but he had refined a tool allowing for the discovery of relational links between zoological groups. His great originality consisted of using embryology to complete his observations of adult specimens.

In 1807 imperial military exercises in Portugal offered him a new opportunity for research abroad. Under the pretext of bringing a collection for the museum back from Lisbon, he left to make a scientific inventory of the country. Crossing an insurgent Spain nearly cost him his life. In Lisbon he made contact with those responsible for collections, organized the material, established exchanges. He was so slow in returning that when he finally departed, military fortunes had changed: England had once again claimed victory, and the Alexandrian adventure had repeated itself! In the end, Geoffroy abandoned his personal luggage and departed for France with his collections.

Figure 2.5. Hunting the giraffe, a more realistic view.
© Aynaud Collection.

In 1809 Geoffroy was named professor of zoology at the Sorbonne in Paris. Between his research, his teaching, and his family, the professor was kept very busy. He would begin working early from bed, by lamplight. At 7 a.m. he would arise and dedicate his day to teaching, dissections, experiments, and administration; in the evening, he tried to find time for enjoyment. He was passionate about literature; his friends included historian Jules Michelet and writer George Sand. He particularly enjoyed comic opera and declared that he was truly impressed by violinist Paganini. In a salon, he knew how to relax the atmosphere and amuse guests with his good-natured banter.

In 1815 the electorate in Étampes awarded him a seat in the Chamber of Representatives. As with every assigned mission, he courageously performed his duty. Geoffroy honored his mandate until the dissolution of the Chamber, after the battle of Waterloo. But he did not continue with politics.

Geoffroy's work essentially addressed comparative anatomy, embryos, and fossils, which allowed him to find links between extinct and living species. He was therefore one of the precursors of evolutionary theory. The Restoration did not change his career in any perceptible way. Geoffroy was a recognized scholar who also had the joy of seeing his brilliant son become his assistant in the 1820s.

In 1825 an initial controversy erupted between Geoffroy Saint-Hilaire and Cuvier surrounding an affair regarding crocodiles in Normandy. Geoffroy argued that the *teleausorus* was a product of the combination of mammal and

CHAPTER 2

reptile, while his colleague described it rather as a crocodile with elongated snout (gavial). Following this dispute, Geoffroy became interested in embryology and the process of monster formation (teratology) in order to explain evolution. But when he wanted to try to maintain chicken embryos in a reptilian stage, Cuvier forbade his experiments.

When the giraffe Zarafa set foot on French soil, the museum's eminent professors were dutifully called in. And in 1827, just as in Egypt thirty years earlier, it was Geoffroy and not Cuvier who would accept the mission, albeit less exotic, of conveying the astonishing animal from Marseille to Paris.

AN OPPORTUNISTIC DIPLOMAT: BERNARDINO DROVETTI

Born on January 7, 1776, near Turin, Bernardino Drovetti (figure 2.6) was successful in law school but, at age twenty, decided to enroll in the French army and participate in the first Italian campaign. Thanks to his military exploits, he quickly moved up in rank. With a return to peacetime, his juridical competence served him when he was named commissioner of the French government in Piedmont. After Napoleon's return from Egypt, during the second Italian campaign, Drovetti continued to rise through the ranks. With peace restored, he became a judge at the new criminal court in Turin. In 1802, the emperor named Drovetti subcommissioner of commercial trade in Alexandria, charged with the reestablishment of relations with Egypt, which had broken down since the departure of the Oriental Expeditionary Force.[3]

With no prior experience in Egypt, Drovetti landed in Alexandria on June 2, 1803, in a country that found itself in ruin and prey to civil war. The situation was so explosive that after risking his life on several occasions, the commissioner thought of resigning. But he soon had the merit of recognizing in Muhammad Ali a future leader and ably gained his esteem. Thanks to his combined talents of former soldier and fine diplomat, and with the support of the new Egyptian ruler, Drovetti managed to thwart the British expedition of 1807. Despite his successes, Drovetti concluded that "the time spent in this country offers only unpleasantness, annoyances, revolution and pestilence, in fact, every scourge by which humanity may be threatened."[4]

In December 1814, following the arrival of Louis XVIII, the notable Bonapartist Drovetti found his position revoked. Not unhappy with his departure from public service and still benefiting from the goodwill of the pasha, the ex-consul decided to stay in Egypt, both to pay off accumulated debt and to un-

Figure 2.6. Drovetti: An opportunistic diplomat.

dertake necessary preparations for his intended marriage to a French resident. For financial reasons, he opened a small trading establishment in Alexandria and pursued his research on Pharaonic antiquities. As early as 1807 he had in fact begun to put together a collection of curiosities, forming collections he would then resell according to his financial needs. Henceforth, he prioritized this self-subsidized, lucrative activity, traveling across the country himself or commissioning agents to work on his behalf. Egypt had been in fashion in Europe since scenes of the *Voyage dans la Basse et la Haute Égypte* (Journey through Upper and Lower Egypt), by painter Vivant Denon, were circulated and several volumes of the text *Description de l'Egypte* (Description of Egypt) were published. Drovetti intended to profit handsomely from this new fascination by selling his collections to European powers.

Another aspect of his commercial trading consisted of furnishing European courts and zoological gardens with exotic animals. In his voluminous correspondence, we note within the space of twenty years the order and delivery of four gazelles sent to Queen Caroline in Naples, sister of the emperor and wife of King Murat; horses from Sudan and an oryx antelope sent to the king of Sardinia; Arab horses to the king of Würtemberg; purebred stallions sent to the court in Vienna; ostrich feathers for the wife of the ambassador of France to Constantinople; Nubian sheep for Count Romanzoff, chancellor of Russia; insects for the Count of Laveau, of the Imperial Society of Naturalists in Moscow; an African

CHAPTER 2

elephant for Charles Felix of Sardinia, Duke of Savoy; shells and fossils for Professor Cuvier of the Museum of Paris.[5]

With his improved financial situation, Drovetti found himself once more consul of France in Egypt. However, as a native of Piedmont, a region that had become independent after the downfall of the empire, he could not remain in this post. But that didn't stop him! Thanks to powerful French support, he obtained his naturalization in 1819 and was accorded the Legion of Honor as well. On June 20, 1821, the French government appointed him to the high post of *Consul général de Sa Majesté le Roi Très-Chrétien dans la vallée du Nil* (Consul General of His Very Christian Majesty the King in the Nile Valley).[6] From then on, Drovetti grew closer to Muhammad Ali and his son Ibrahim, reinforcing his relationship through well-timed gifts. It was during this period that the pasha launched his modernization project by hosting French technicians in Egypt.

In 1822 Champollion graced the world with his decoding of a fascinating civilization. The first volume of his *Description of Egypt* was published in 1809 and the twenty-third appeared in 1838: this great work dazzlingly demonstrated an immense variety of areas of interest so attractive to scientists; his writings contributed to the creation of a mythical Egypt, still present within the French national unconscious. Beginning at that time, Egyptophilia gained considerable importance in France, combining political interests, artistic sensibility, and scientific preoccupations. Louis XVIII was not indifferent to this fascination. During his reign, the holdings of the *Muséum central des Arts* (Central Museum of the Arts) were enriched by important sculptures. However, the king became suspicious due to the actions of certain opportunists, so he refused to buy Drovetti's first collection, at a price he found to be excessive. Unofficially, the clergy deemed sacrilegious the idea of importing objects purported to be dated prior to the creation of man according to Genesis. The collection would finally be acquired for four hundred thousand lire by the king of Sardinia, who took advantage of it in order to found the first museum of Egyptology in Turin.

Even though he strove to bring together France and Egypt, Drovetti would be considered alternately to be a plunderer, an unscrupulous adventurer, and a relentless promoter of Egyptian treasures. Champollion was ambivalent toward the consul, thanking him as a source of priceless information and a door-opener to ancient splendors, all the while reproaching him for having participated in the "trade of ancient things."[7] During his trip in 1828, the scientist even suspected that Champollion sought to prevent his excavation projects. Champollion's judgment against the consul's actions was harsh: "I do not appreciate his political leanings and conduct in Egypt where he has concerned himself only with his interests allied with those of the pasha, without paying the least bit of attention to those nationals he had been paid to protect." But in 1824, with the arrival of Charles X to the throne, the consul-adventurer Drovetti foresaw a new opportunity within his grasp.

3

THE PASHA'S GIFT
Diplomacy at Stake

The year 1824 bore witness to the Egyptian involvement in Greece, the enthronement of Charles X in France, and the project of sending an Egyptian giraffe to the French king. These three events are directly or indirectly linked.

GREECE: REBEL AND MARTYR

In 1815 some Greeks residing in Odessa, Russia, founded a secret society to prepare a Christian revolt in the Balkans. The first two attempts failed, while the third, under the direction of Bishop Germanos, inflamed the entire peninsula. In 1821 massacres were perpetrated by Greeks in Tripolitsa. The next year, the Turkish response was merciless in Constantinople and in Chios. At a congress in Epidaurus, the Greeks proclaimed their independence, and the revolt was organized there and abroad. Philhellenic committees were constituted in Paris, London, Germany, and even the United States.

Having set sail from England by his own means in July 1823, Lord Byron did not find much in Greece to affirm the romantic ideal he had envisioned of this war for independence. Instead, he saw discord, aggression, and fraud reigning there. The English poet deemed the people to be brave but undisciplined, the rebels, whose leaders were divided and corrupt, as cruel and turbulent. He attempted to organize the different factions but died on April 19, 1824, at age thirty-six, of a fever contracted during one of his daily outings on horseback. Greece then declared a national time of mourning, and his body was returned to Great Britain. Several paintings recognize Byron's engagement: *The Oath of*

CHAPTER 3

Lord Byron in Missolonghi, by Lipparini (1824), and *Lord Byron on His Death-Bed*, by Joseph-Denis Odevaere (1826).

France was hostile to Turkish brutality, even more so because Egyptian soldiers sent as reinforcements to the sultan had been trained by former officers of Napoleon's armies. Their military chief of staff was also a Frenchman, one Colonel Sève. The same year, Eugène Delacroix painted *The Massacre at Chios: Greek Families Awaiting Death or Slavery* (figure 3.1). Exhibited in a salon, the painting was immediately considered to be a manifesto of new Romantic painting, due to its resolutely modern subject matter as well as its artistic style. Delacroix shows an unjust and unequal combat between armed warriors on horseback, and a population composed largely of women and children. The painting is full of characters, saturated with detail, all portrayed in violent colors and disorganized lines that contribute to an impression of chaos. The painting is uncontestably political, judged to be either a work destined for the public good or a work of propaganda, depending on one's point of view.

Figure 3.1. *The Massacre at Chios*, by Eugène Delacroix, 1824.
Louvre Museum, Paris, Public Domain.

THE PASHA'S GIFT

On the Ottoman side, Muhammad Ali offered his support, as he served as vassal to Sultan Mahmoud. His son Ibrahim crossed the Gulf of Corinth and disembarked in Morea with twenty thousand men. The war redoubled in fury, and the country was devastated. In early 1826 Ibrahim received reinforcements of artillery and supplies. For the city of Missolonghi under siege, the situation was desperate. After a year of tenacious resistance, the Greek leaders developed a plan to escape from the city. But in the panic that ensued, only one thousand of the nine thousand inhabitants managed to flee. On the following morning, Palm Sunday, Turks entered the city, and most of the remaining Greeks preferred to blow themselves up with cannon powder rather than to surrender. The survivors were massacred or sold into slavery; the Turks placed some three thousand heads upon the ramparts.

After that event, which did not prevent the sultan from reconquering the Peloponnese, sympathy for the Greek cause grew stronger in Western Europe. Delacroix painted a second canvas that he titled *Greece on the Ruins of Missolonghi* (1826). The artist no longer portrayed the massacres; the ruins are scarcely perceivable in the form of rubble in the foreground. A woman is depicted as an allegory of Greece, akin to *Liberty Leading the People* (1830), Delacroix's most famous work. Inspired by one of Byron's odes, the painter depicted Greece with the traits of a young woman in traditional costume. Her chest uncovered and arms outstretched, almost kneeling on the remains of a martyred city, she represented a living condemnation of the violence unleashed upon Greece in its revolt, and a triumphant symbol of the next resurrection of this nation. The painting was exhibited in 1826, during a demonstration organized by the Lebrun Gallery, in order to raise public funds to help the Greek insurgents.

Music also played a role in the fight. The composer Rossini launched his *Siege of Corinth* on October 9, 1826, in the theater of the *Académie royale de musique*, with the Italian libretto written by Luigi Balocchi and Alexandre Soumet. Thus spoke Cléomene, leader of the Greeks:

Depuis long-temps du vainqueur de Bysance,
Qui de toutes parts
Assiège nos remparts,
Nous avons affronté la farouche arrogance.
Amis! votre vaillance
Chaque jour du tyran sait braver la fureur;
Mais l'avenir m'effraie. . . . Hélas! au champ d'honneur,
Nos plus braves guerriers trouvent leurs funérailles;

CHAPTER 3

Des fléaux dévorants assiègent nos murailles....
Le glaive musulman, le bronze des batailles,
Moissonnent à l'envi le peuple et les soldats;
Maître de nos états,
Mahomet furieux nous menace et nous presse;
Des flots de sang vont inonder la Grèce....
Pour fuir le joug du tyran,
Ô ciel! quel parti prendre?
Faut-il combattre encore, ou bien faut-il se rendre?
Ô terrible moment!
Le danger est extrême.

[For such a long time from the conqueror of Byzantium,
Who from every direction calls,
Has laid siege to our rampart walls,
We have confronted unbridled arrogance.
Friends! Your valor hence
Each day knows how to brave the tyrant's fury;
But the future is a fright to me... Alas! on the field of honor see
Our bravest warriors confront their final hours;
Devastating scourges lay siege to our walls and towers....
The swords of Muslim gladiators, the bronze of battle's powers,
Are amassed by people and soldiers time and again;
Master of our land,
Furious Muhammad, threatens and pressures us;
The swelling waves of blood will inundate Greece....
To flee the tyrant's yoke,
Oh heaven! What side to take?
Must we fight on, or surrender our fate?
Oh terrible time!
The danger is extreme.]

Berlioz also composed a political work, his *Scène héroïque à grands choeurs et à grand orchestre*, on a text by Humbert Ferrand. "The matter of the Greek revolution occupied every mind at the time," noted the composer in his *Mémoires*. Composed in 1825–26, the piece was performed on May 26, 1828, in the Conservatory room, thanks to the director of the *École des Beaux-Arts*, Sosthène de la Rochefoucauld:

Lève-toi, fils de Sparte! allons!... N'entends-tu pas
Du tombeau de Léonidas
Une voix accuser ta vengeance endormie?
Trop longtemps de tes fers tu bénis l'infamie,
Et sur l'autel impur d'un Moloch effronté
On te vit, le front ceint de mépris et de honte,
Préparer, souriant comme aux jours d'Amathonte,
L'holocauste sanglant de notre liberté.

[Rise up, son of Sparta! Let's go! Don't you hear,
From the tomb of Leonidas near,
In your sluggish vengeance his accusing plea?
For too long by your irons you bless infamy,
And on the impure altar of a shameless Moloch come,
Your forehead encircled in disdain, a disgraceful front,
As you prepare, smiling as in the days of Amathonte,
The bloody holocaust of our freedom.]

Another committed artist, Victor Hugo, published his volume of poems, *Les Orientales*, in 1827 and 1828. While the writer was not known to be among the first Philhellenists, he was yet one of the most ardent and constant among them. The immediate success of this volume ensured his renown as a Romantic poet and allowed him to affirm his style. It was obvious that the war for Greek independence was not subject to indifference in the French political context:

Les Turcs ont passé là. Tout est ruine et deuil.
Chio, l'île des vins, n'est plus qu'un sombre écueil,
Chio, qu'ombrageaient les charmilles,
Chio, qui dans les flots reflétait ses grands bois,
Ses coteaux, ses palais, et le soir quelquefois
Un choeur dansant de jeunes filles.[1]

[The Turks have passed there. All is ruin and grief.
Chios, the island of wine, is but a somber reef,
Chios, that gave shade to the bowers,
Chios, whose waves reflected its forest towers,
Its hillsides and palaces, and in the evening perchance
A chorus of young girls in the dance.]

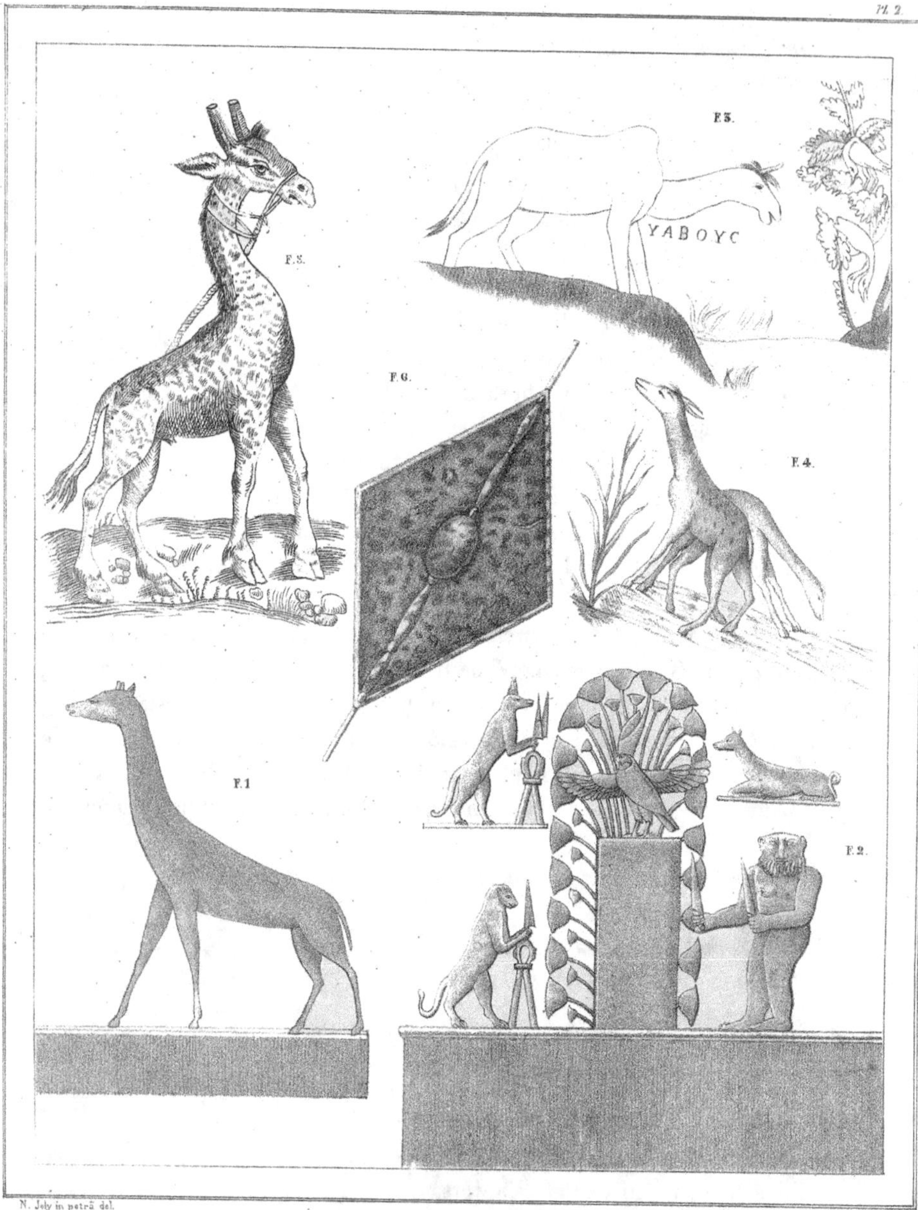

N. Joly in petrâ del.

Lith Belor à Toulouse.

détails relatifs à l'histoire de la Girafe.

Figure 3.2. One standing giraffe and one lying down, at the temple of Hermonthis (according to Jomard, in *Description de l'Égypte*, 1820, 2nd edition, attributed to Panckoucke).

In Joly, N., *Notice sur l'histoire, les mœurs et l'organisation de la girafe*, Toulouse, 1844.

This mobilization of artists surrounding issues as sensitive and romantic as a country's fight for freedom while under a tyrant's yoke helps us to understand the passionate climate surrounding the giraffe's appearance. We can also grasp the pasha's strong interests, as he sought to regain prestige among European powers. We note again that Muhammad Ali, in disagreement with England about Greece, deemed it acceptable to decapitate a few British citizens and display their heads on Cairo's ramparts. Needless to say, the English court did not appreciate this.

In 1825, however, Governor Mouker Bey of Sudan offered two giraffe calves to the Egyptian pasha. The idea of presenting a gift to the French king to enhance his zoo at the *Jardin des Plantes* was then conceived. This news spread to the ears of Henry Salt, English consul, who also claimed a giraffe for his king, George IV.

The *self-serving* gift that was Zarafa, as well as Drovetti's disturbing game, was clearly visible in private correspondence with his friend Balthalon, a merchant from Marseille. In the same letter, in fact, the consul announced the soon-to-be-arriving giraffe and denounced the politics of the French government, his official employer. Drovetti revealed himself to be an unscrupulous merchant thinking only of profit rather than acting as a principled diplomat: "All of our traders on this level, almost entirely young people who have avoided the draft, full of ideas they call liberal, but which greatly resemble Jacobinism, render my position untenable." The next part smacks of high treason: "As long as I had only to fight against French antagonism, I felt at least occasional satisfaction in seeing them humiliated, but when unrelenting opposition from some Frenchmen appeared, those who, through Jacobinism, philhellenism, and who knows what else, take pleasure in vexing me and dousing me with repugnance, this is a battle I do not wish to fight, and so retreat is the wisest course of action, and the most prudent." Fair enough, but Drovetti was still not ready to leave his post.

A REACTIONARY MONARCH: CHARLES X

Born in 1757, Charles X (figure 3.3) was sixty-seven years old when he acceded to the French throne upon the death of his brother, Louis XVIII. His religious devotion disconcerted the Parisians, who were willfully anticlerical. The appearance of the new king at the funeral of the past king, dressed in purple, the color of grief associated with French kings, launched a rumor that he would become a bishop! Caricatures showed him celebrating mass before members of his family. In 1805, after the death of the last of his mistresses, Louise de Polastron,

Figure 3.3. *Portrait of Charles X in coronation costume*, by Jean Pierre Marie Jazet, engraver. Print on vellum paper, 1825.

© Société Chateaubriand, Domaine départemental de la Vallée-aux-Loups, Maison de Chateaubriand, inv. GE.961.324.

his conduct became irreproachable, and his piety fed the idea that he was an instrument of the clergy.

It is worth noting that Charles X rekindled tradition by organizing his coronation at the cathedral in Reims. This royal act led his subjects to fear the worst. Although he had declared his acceptance of the Charter, he could not resign himself to the role of a constitutional king, preferring, he said, "to saw wood rather than to reign in the fashion of the king of England." Despite a few liberal actions taken early on, he showed himself faithful to his absolutist positions, choosing as his minister a certain Villèle, whose authoritarian politics would strengthen opposition movements. From 1825 to 1827, his ignorance of political reality led him to take a series of unpopular measures, such as the law referred to as *le milliard des émigrés*, which accorded 33 million French francs of annual income to emigrated nobles whose assets had been seized during the Revolution.

In addition, aligning himself with the current trend, the king was passionate about antiquities. Due to good advice from Champollion, he bought in succession three of the large collections then for sale in Europe–those of Durant, of Drovetti, and of Salt, the consul's English rival. All in all, consisting of around nine thousand pieces, these acquisitions would constitute an essential part of the Louvre's Egyptian collection. By royal decree on May 15, 1826, Champollion was appointed curator of a new department, inaugurated December 15, 1827. Between these two dates, a "beautiful Egyptian female" would enter the French scene.

Written in 1824 by the administration of the *Muséum royal d'Histoire naturelle*, "Instructions for travelers and for employers in the colonies"[2] urged the latter to facilitate the introduction of exotic animals in order to enrich the royal collection. As soon as this circular was distributed, Drovetti hurried to send the king a couple of antelopes. But he thought he could do more. . . . Which one of them, pasha or consul, first conceived the idea of offering a giraffe to the king of France?[3] This is unknown. Making the gift was, however, in the interests of both of them. Moreover, for a monarch as contested as Charles X, it would certainly be advantageous for him to impress his subjects with some exotic marvel.

A PRECIOUS PACKAGE

On August 2, 1824, in his correspondence, Drovetti mentioned Zarafa's forthcoming arrival. Frustrated by the difficulties of his position, which worsened his health, the consul thought about resigning and leaving Egypt. In addressing his friend Pierre Balthalon, he evoked a certain Parisian, visibly near the new power:

CHAPTER 3

Please communicate this item to the friend in Paris; you can assure him that the Basha [sic] has given orders to acquire a giraffe he wishes to offer as a present to the king. If it arrives safe and sound, it will have cost quite a considerable sum. To sustain the animal, it must be taken when young and brought here from Ethiopia with an envoy of cows to furnish the milk necessary for its nourishment. As for the rest, all relations that France could or would like to maintain with the Basha [sic] or Egypt are no longer of interest to me and I will be performing strictly minimal duty here as I await my leave.[4]

Zarafa, the first living giraffe to set foot in France, as affirmed by the official Notice, "is a female, of five or six moons (around six months old) at the time of capture: with her was another giraffe of the same sex and age; the desert Arabs who took them sold the two of them to Mouker Bey, governor of Sennar, who sent them as a present to the pasha of Egypt, his superior: the latter kept them for three months.... This one has made the trip from Sennar to Cairo, partly by walking with caravans and partly by descending the Nile in a boat designed especially for her."[5]

The desert hunters knew that such an animal had to be captured before being weaned in order to have a chance at survival in captivity and adapting to domesticity. The trick was to add animals to enable it to nurse; this function would be served first by camels and then by Egyptian cows. Once the giraffe calves were captured, the regional governor took good care to fasten the living packages to the backs of camels, before beginning the eight- to ten-day trip by caravan needed to reach the next city of Sennar. From there the captives were embarked on a felucca; they descended the Blue Nile to Khartoum and then the length of the Valley of the Kings to Alexandria, all within sixteen months.[6] "It's only when they are nursing that one can take hold of them and tame them; even then they often managed to twist and dislocate their neck to get loose from their ties: at other times, they would refuse all food and would die of hunger. But if one managed to keep them without incident for several days, they calmed down, got used to the situation and followed horses, camels and their keepers without a leash."[7]

To protect them during the trip, the pasha's representative in Cairo knotted around their neck an amulet that was to have touched the venerated tomb of Sayyidah Zaynab and that contained a Koranic verse intended to protect them from the evil eye.[8]

Consul Salt claimed one of the giraffe calves for the king of England. As it happened, one of the animals appeared to be of more fragile health, and so the undertaking took on a high political risk. It was essential for Muhammad Ali to manage the sensitivities of the rival powers. Finally, they drew straws to decide which giraffe would go to Paris and which one to London. Knowing the rivalry that opposed Drovetti and his English counterpart, as well as the privileged relationship of the former with the pasha, one might wonder as to the actual result of this random draw.... Whatever the case, Drovetti sent an immediate and triumphant dispatch to the French court: "I am most pleased to announce to Your Excellency that fate has been favorable to us. Our giraffe is in fact strong and vigorous; the one befallen the king of England is unhealthy and will not live long."[9]

In fact, the giraffe in England would die two years after its arrival in Windsor.

What follows is a detailed account of the fabulous destiny of Zarafa, the first giraffe to set foot in France.

4

ZARAFA IN FRANCE
A Royal Triumph

September 29, 1826: Consul Drovetti wrote from Alexandria to Mr. Bottu, foreign affairs agent in Marseille, to provide instructions regarding care of the precious traveler, who, like all foreign passengers, would comply with the rules of quarantine upon arrival:

> I also ask you to ensure that the Giraffe will be given everything she needs with regard to nourishment, and placed in lodging with a comfortable temperature in Lazaret. Although this quadruped may not be as sensitive to the cold as her native country's latitude might lead one to believe, I think she should remain in Marseille until the warm season arrives.
>
> If the cows I am sending for the Giraffe don't provide enough milk for her after their arrival, I would be grateful if you would advise the person designated as responsible for that nourishment to procure twenty-to-twenty-five liters a day for her. An ongoing supply of milk is essential: one Giraffe, sent three years ago to the noble lord in Constantinople, perished because they stopped giving her milk to drink in order to save money.
>
> I also take the liberty to place in your good care, two antelopes that I am sending to the King, I have alerted the Minister of his House.
>
> The male is affected with an illness that, while skin-related, seems to have an internal origin. Scabies is presumed unlikely, for this would have already been contracted by the female.
>
> I will write to His Excellency, Minister of the King's House, to have my black servant (a keeper responsible for the animals' journey) sent to Paris and to have him stay in the vicinity of the Giraffe, if deemed opportune, as this animal could have difficulty adjusting to a European handler.

CHAPTER 4

While the consul took the risk of sending an ailing antelope to the king, he took all precautions with regard to the giraffe, sending his own servant to accompany her. In 1823, Hassan, the man in charge of Drovetti's stable, had escorted the first giraffe, destined for the sultan of Constantinople, and could personally vouch as an eyewitness to the reasons for that animal's death.

September 30, 1826: The already rather large animal was led onto a brigantine, a small sailing vessel with one deck and two masts. It would sail from Alexandria to Marseille, entrusted to Drovetti's nephew and four Arab grooms, who were to watch over the animal as though it were their prized possession. The cargo comprised a veritable Noah's ark, including "the four horses of the general-baron Pierre Boyer, two asses, three milk cows, four sheep, two lambs, two antelopes given by Drovetti as a personal gift, all destined for the Royal Zoo, plus one antique statue."[1]

Zarafa's boat also probably transported the French officer. At least it is known that he returned to Marseille during the same period. And history is full of coincidences! A veteran of the Egyptian campaign, the general Pierre François Xavier Boyer had been recruited by Muhammad Ali to train his army. But after only two years, he resigned from his duty.... As a double agent, it is thought that he was secretly charged by France to dissuade the pasha from attacking Greece and to turn against the Turkish oppressor instead. Once his intentions were discovered, he could not continue. And so, in one maneuver, the pasha sent back the king's puppet and set in place a new queen: Zarafa.

What is certain is that Drovetti resented Boyer: "I can't begin to tell you of all the hassles General Boyer and his companions have caused me; they left a few days ago for Marseille. Since their departure, our position has changed much in Egypt due to their continuous state of discord, and for the first time in twenty-three years I am seeing our rivals triumph, those to whom these men had given the weapons to fight against us."[2]

Captain Manara, commander of the brigantine sporting a Sardinian flag, agreed to the sum of 4,600 francs to divert from its regular line of Alexandria–Livourne. Zarafa's maritime crossing resembled a farcical exploit. The giraffe's neck emerged onto the ship's deck through an open panel whose edges were lined with soft material to protect her from the lurches of the rolling boat. A canopy was placed above her—an oiled canvas stretched between four posts—as protection from the sun, rain, and sea spray.

For maximal security, one of the king's corvettes, the *Echo*, was charged with escorting the brigantine to the right port. Scarcely one week after their departure from Alexandria, the two vessels became separated during a violent storm.

Nevertheless, during the three-week crossing, the Mediterranean was merciful to the quadruped, who disembarked safe and sound.[3]

October 23, 1826: On that day, the captain's declaration was listed in the registers of the port in Marseille: "Captain Stephano Manara, Sardinian, of the brigantine *Les Deux Frères*, weighing 255 tons, with a total crew of thirteen persons, having departed from Alexandria September 30.... He remitted his permit and those of his passengers, noting as well four Arab animal handlers with no permit."

The prefect had just published his *Statistique du département des Bouches-du-Rhône*, which outlined in detail the special rules Zarafa's boat had to comply with:

> Property arriving from the posts of the Levant (Eastern Mediterranean) or from Barbaria (North Africa) with clear permit and without ill or deceased, received in the port of Pomègues ... must report to the Registry office to make any oath or declaration that the law requires of the one in charge, and to respond to any questions from health officials. Among the ships, some are admitted without restriction; others are submitted to what is called quarantine for observation and placed in the port's custody; a public health officer is posted to them, and they are, in addition, surveilled by health officials assigned for this purpose.

The giraffe, her keepers, and companions for the journey began therefore by staying on board for reasons of public health and also to take care of some administrative difficulties. The minister of foreign affairs refused to spend a dime on any animal disembarked upon French soil, even a royal gift. And so they awaited a green light from the relevant authorities. Moreover, the customs office was still seeking reimbursement from the prefect of "fees incurred by Lord François Bonnet that the barge *La Truite* had recently brought from Alexandria, and for the wild feline that the same ship disembarked in the Quarantine station," in the amount of eighty-eight francs demanded since June 8 of that year.

October 26, 1826: Mr. Bottu, foreign affairs agent posted to Marseille, wrote to the Count of Villeneuve-Bargemon, prefect of the Bouches-du-Rhône, to announce the arrival of the boat with giraffe on board, and specifically to transmit instructions regarding the animal:

> My Lord Count,
> I have the honor of informing you that the Sardinian brigantine, *I Due Fratelli*, commanded by Captain Stefano Manara, arrived in the port three days ago, transporting from Egypt several animals destined for the King's zoo, namely:

CHAPTER 4

– A giraffe sent by the Pasha, with three cows assigned to furnish necessary milk for this quadruped;
– Two antelopes, male and female, sent by the Consul General Sir Drovetti, who placed them all in the safekeeping of one of his groomsmen, and one black servant.
– In accordance with the order expressed in a dispatch from His Excellency the Minister of Foreign Affairs, dated the tenth of this month, whose excerpt you will find attached, my lord, all care and necessary fees for this type of delivery, from the moment of arrival in France, are ascribed to the superior administrative authorities, for whom I must record them; His Excellency reserves for his department only payment for transport fees and other related expenses incurred outside the kingdom.
– Moreover, I think it appropriate, my lord, to attach to the present missive an excerpt of Consul Drovetti's letter, containing several observations relative to the care to be provided to the various animals that I believe should urgently disembark.

I take this occasion to renew my expression of high esteem and respectful consideration for you, and for whom I have the honor to be,
My Lord Count,
Your very humble and very obedient servant.

October 27, 1826: The following day, nine public health officials in Marseille, knowing that Mr. Bottu was not responsible for expenses incurred, officially asked the prefect to authorize an advance for the fees the animals would surely cost the quarantine station. How much would the care of a giraffe cost? For the first month of quarantine, an account was stringently established. Supplies tallied were related to the feeding and care of one giraffe, two antelopes, and three cows, in the following amounts:

Food for two animal guardians for thirty-one days at 3 F per day	= 93.00 F
Wages of above guardians in the amount of 30 F for one and 20 F for the other	= 50.00 F
Herdsman salary at 1 F per day	= 30.00 F
755 kg hay	=107.30 F
735 kg straw	= 59.40 F
5 hectoliters beans	= 101.25 F
750 kg bran	= 45.00 F
4 hectoliters, 4 decaliters of barley	= 72.00 F
Supplies according to the coachman's tally	= 40.50 F
	598.45 F

October 28, 1826: Even before receiving the Officials' request, the prefect urged them to take the greatest care of the unloaded animals, especially the Giraffe, repeating almost word for word Drovetti's instructions regarding the quadruped's care.

October 31, 1826: A week after her arrival, Zarafa finally disembarked, but without setting foot in Marseille's port. A fishing boat and its crew transferred the whole load of animals from the brigantine to the lazaret (figure 4.1). As a precaution, giraffe and antelopes were equipped with blankets made of tartan wool.

A painful isolation began. But one cannot contest quarantine: in the eighteenth century, in the space of two years, the plague had carried off more than half the population of Marseille, then called *the dead city*. Since then, port authorities had applied draconian health measures, especially to combat the spread of yellow fever, which risked affecting passengers coming from South America via Spain. The ships had to set anchor five kilometers off the coast and undergo a period of isolation. Passengers were then admitted for several weeks into an infirmary linked to health services in Marseille or transferred by small boat to the famous lazaret for quarantine.

Figure 4.1. Lazaret (quarantine station) in Marseille.
© Aynaud Collection.

CHAPTER 4

The place had a terrible reputation. In 1840, Clot-Bey, surgeon in Marseille who had become director of Egypt's health services and founder of the first medical school in East Asia, gave a harsh description of it:

> There is real danger going there and stopping there.... The passageways are in a state of total decrepitude... the unhappy ones in quarantine are lodged in real hovels.... These are *terrible prisons* where travelers already fatigued by the crossing are at risk of falling ill.

As for the officials,

> they know how to exploit this kind of prison like a fertile provincial meadow they have inherited, with good returns.[4]

Five months before Zarafa's arrival, a young Arab man hadn't formed such a harsh impression of it, but rather one to the contrary. On *La Truite*, accompanied by the *wild feline* who had caused so many problems for the prefect, a delegation of fifty-four learned people and scholars were sent by the pasha to Europe. One of them, Rif'â-at-Tahtawi, gave a detailed description of the lazaret: "The compound consisted of a palace, a garden, and some buildings, all very solid. It was there that we noted the quality of this country's buildings. They were constructed with care and planned with many parks, ponds, etc."

The young man was especially struck by the strangeness of Western customs:

> For example, we saw several French servants arrive, whose language we didn't know, and who brought us nearly a hundred seats for us to sit in, for in this country, they are stupefied to think that we might sit upon the ground. Then, at mealtime, the servants set these seats around the table, one per guest. After which, the dishes were passed out to us, each one receiving in a small individual plate something that we had to first cut with a knife set before us, then bring to our mouths with a fork, and not with our fingers. Here, no one eats with their fingers, just as no one uses the fork, knife or glass of another guest: they assert that it is cleaner this way and more courteous.[5]

Situated three hundred meters to the north of the city, on an area of 230 kilometers, "the Lazaret of Marseille occupied such a great expanse that it became, by fact and by law, the only large establishment of this type that existed on the Mediterranean French coastline." Before Zarafa was stationed there in quarantine, several famous historical figures had also spent time there. In 1802 the ten thousand men of the Napoleonic army upon return from Egypt were sequestered there, in the new enclosure: "We note there are no trees or shrubs in this vast space; this is so that no bird may perch there, and one takes great care to keep them away for fear that, by taking bits of cotton or wool, they could transport some contagious germs into the city or countryside."

The Grand Enclosure could offer more comfort and host more prestigious guests: it was there that "on May 21, 1816, Her Royal Highness *Madame la Duchesse* de Berry, arriving from Naples, set an example for obedience to health laws, and spent fifteen days in the lazaret."[6]

Zarafa would spend seventeen days there, under high surveillance. The health officials ensured her supply of milk, even though one of the cows had stopped providing milk during the crossing. Fresh hay and straw for her bed were provided. They kept watch on her in shifts. As for the foreign animal keepers, they were only allowed a frugal diet, one candle, some oil for their lamp, and a pile of wood for heating. The account of November 14 noted a total invoice of 540.40 francs. The officials posted it to the prefect on November 16, recalling to his attention the balance still outstanding.

Solicited and pressured from every direction, it was then that a crucial person appeared, one who would become central to Zarafa's protection: Christophe de Villeneuve-Bargemon, prefect of the Bouches-du-Rhône (figure 4.2). Born in 1771 in Bargemon, to the north of Draguignan, the Count of Villeneuve hailed from a Provençal family of old lineage. After 1789 he remained faithful to the king, serving in the Constitutional Guard. The protection of General Lacuée allowed him to occupy several administrative posts in the consulate. Between 1806 and 1815, he became prefect of Lot-et-Garonne, and then of Bouches-du-Rhône in 1815, with the title of state advisor. He was certainly known to the new

Figure 4.2. "Christophe de Villeneuve-Bargemon, préfet des Bouches-du-Rhône en 1826."
© Château d'Arlay Collection.

CHAPTER 4

king, as the count was an active member of a secret society, founded in 1810, the *Chevaliers de la Foi* (Knights of the Faith), whose confidants had to demonstrate Christian virtues and favor the reestablishment of royal legitimacy.

A learned and zealous public servant, Villeneuve was at the origin of many municipal building projects, including Arles's theater restoration and the construction of the *Arc de triomphe* of the *Porte d'Aix*, in Marseille. He was active in the *Académie des Sciences* in Marseille and in the creation of the *Société de statistique*. His name remains associated with the *Statistique du département des Bouches-du-Rhône* (with an atlas dedicated to the king), whose four volumes appeared between 1821 and 1829. With a passion for biology, the prefect took particular interest in his "adoptive daughter." Even before informing the Minister of the Interior, he took the initiative of constructing a stable in a prefecture courtyard and continued to take steps to ensure the well-being and safety of his boarder.

November 8, 1826: The prefect wrote to the Minister of the Interior. He gave him news of the giraffe, then in quarantine, and informed him of his plans for the animal's stay:

> Awaiting your arrangements regarding expenses, everything leads me to believe that they are of some importance here, and this is what I wish to convey. The Giraffe must be placed in a lodging of a suitable temperature. The Consul general has observed that while the quadruped is not as sensitive to the cold as might be expected considering the latitude of her native country, it will be necessary to leave the animal in Marseille until the warmer season has arrived. No premises among those in the vicinity of the prefecture and which up till now have served to receive animals sent from abroad are suitable for use in this circumstance, especially for a long stay. Another site must be found and made available.
>
> Regarding all these considerations, to which are added the certainty of the upcoming departure from the lazaret (to occur on the fifteenth of this month), Your Excellency will ascertain the importance of sending me instructions as soon as possible.
>
> P.S.: You can assure the administrators of the Natural History Faculty that I will take special care to preserve this precious animal. I have already confirmed a space large enough with good noonday exposure. It will even be provided at no cost. I will have built a vast shed of wood planks, that we will heat by means of straw matting. If the seasonal cold renders it necessary, this construction will require some funds; one must as well provide for the animal's food, and for some companion cows that supply her with milk; the keepers must also be provided for on a daily basis. Progress will be made under my direction, in the interest of saving as much as possible, but it is essential that you have some funds sent to me at the first opportunity.
>
> P.S. Mr. Drovetti's nephew has just noted that the prefecture hangars are very suitable for the care of these animals.

November 11, 1826: On that Saturday, the animals from Alexandria had to be designated "free to circulate." But the prefecture's stable wasn't finished, so they remained in the lazaret until the following Tuesday. Then Villeneuve wrote to the customs director not to interfere with the transfer of the animals, and he informed the public health officials, "Once everything has been prepared, we will go to the lazaret to take delivery of them during the course of the day, all except for the Giraffe; for various reasons I think it more suitable to bring her into the city at a much later hour. Consequently, please authorize the captain of the lazaret as soon as possible to open the door between 10:00–11:00 p.m. to release the Giraffe, upon request by authorized agents responsible for this action."

November 14, 1826: The "various reasons" that raised concerns for the prefect included the public rumor concerning a legendary monster, a dragon-like hybrid beast–the *Tarasque* incarnate–that the crowd might harass out of curiosity, or in the worst case, tear to pieces in a movement of panic. The shelter, built in the prefecture's courtyard was so large, its use so anticipated, that people expected the imminent arrival of a surely gigantic and dangerous beast. The authorities feared some misbehavior. That was why the cows and antelopes were released from the lazaret first, at midday and in the midst of a row of onlookers. That very night, under the guard of her keepers and flanked by an escort of policemen on horseback, Zarafa would extend her stroll into the streets of Marseille (figure 4.3).

An observer provided a detailed account of this secret convoy's walk, which caused some local authorities to break into a cold sweat. In one narrow alleyway, Zarafa stopped dead in her tracks, no longer seeing the mounted policeman leading her, but fortunately she regained her course:

> As soon as the giraffe once again saw that horse she had suddenly lost sight of, she became calm and walked very closely behind, along with the Arabs who held her with four ropes; but the horse was unsteady, and his rider had difficulty controlling him. The horse couldn't stand the fact that the giraffe occasionally sniffed his hindquarters. She had to cross several public promenades, and she continually sought to reach the branches of nearby trees, without losing sight of the horse she had chosen as guide, that she followed faithfully right up to the stable prepared for her.

From the Museum of Natural History, the veterinarian Bosc sent his recommendations, imagining the worst:

> This animal has special need for a clean, aerated, and warm stable, and it would be desirable for it to have at its disposal a space to go during good weather.

Figure 4.3. *La Bête au bois dormant*. Text and illustrations by Robida. The title is a play on the seventeenth-century fairy tale by Charles Perrault, *La Belle au bois dormant* [*Sleeping Beauty*]; bête means "beast."
© Aynaud Collection.

One must be particularly vigilant regarding cleanliness of the area around it, and that its keeper not be negligent in its care in any way.

As for its nourishment, the keeper will follow instructions given in Egypt for that, and what he observes to best suit the animal.

If this giraffe were to die, the hide should be removed with the utmost of care; the flesh covering the bones should be removed likewise, and the hide and the skeleton sent, taking precautions for their preservation during the trip.

November 18, 1826: The prefect rendered an account to the Minister of the Interior regarding transportation of the giraffe into its tailor-made stable, her health, and all related expenses:

My Lord,

I have just transferred the Giraffe and the two antelopes into the constructed sheds in one of the Prefecture courtyards, as I had relayed to you their imminent release from quarantine in my letter dated the 8th of this month.

The Giraffe is very beautiful and has really started to regain her initial energy since arrival in the Lazaret, as the sea crossing had somewhat perturbed her. I have fittingly equipped the premises where she is currently placed, especially regarding the necessary temperature, and the most assiduous care is given to her by the two Egyptians who accompanied her. Her nourishment, which I specified in a preceding letter, consisting mainly of a rather large quantity of milk, demands the maintenance of three cows that made the trip with her. The expense, including that of the keeper's wages and food for the antelopes, amounts to about twenty francs per day. Regarding fees for disembarking and for quarantine, I will not know the total amount until I receive the bill from the Public Health officials. It is therefore indispensable that a sufficient sum be sent to cover these expenses, that will continue to be paid, balancing savings with the necessity of preserving from decline such an interesting item of natural history. It is a female giraffe. Her height to the top of the head is eleven feet, one quarter inch. Artists drawing her have already begun the task on a request received from Germany.

I also think it very important that Your Excellency send me the instructions that the Administrators of the Museum have deemed necessary to provide for the conservation of these animals. I will take care personally to transmit all observations that have been made. We will then occupy ourselves with her transfer to Paris, but as we will not be able to think of moving her before the month of May, we have time to foresee and arrange everything in advance.

P.S. The most detailed measures have been taken for the Giraffe's upkeep. As she is close by, I can keep watch over her at every moment and save the State from spending enormous sums that it would have cost to shelter her elsewhere.

CHAPTER 4

The prefect had in fact thought of everything. Set up and later dismantled for a total cost of fifty-five francs, the stable was a "vast shed of planks heated with straw matting if the rigors of the season required it." Following the advice of the sender Drovetti, who was distrustful of stoves and lamps, it was at first lit by sunlight from two windows and a large paned glass door. The animal's heat was enough to maintain a pleasant temperature, along with two sturdy horses residing on the other side of the partition. Rented for twenty-five francs, the wall hangings served as protection from drafts.

The Egyptian keepers also spent the winter in the shed. Cots and blankets were requisitioned for them from the store of a local military garrison. They received two new lanterns and a supply of oil. A daily allocation including their food was valued at 46.50 francs per month. One of the keepers suspended his hammock at the height of Zarafa's head, to better keep watch over her and caress her between the ossicones. The other milked the nurse cows, which grazed on the prefecture's lands.

November 21, 1826: Méry, journalist for *Le Messager*, foretold a somber future for Zarafa:

> The office in the *Jardin des Plantes* in Paris has been graced with a giraffe ever since the traveler Levaillant, of dubious memory, killed one on the banks of the river Orange in Southern Africa; the cadaver of this gigantic quadruped is in a ruinous state and soon will amount to nothing more than a pile of straw dissipated by the wind. The giraffe from the Egyptian pasha could not come at a better time. She will spend the winter in Marseille, leave for Paris, and expire *en route*; she will be stuffed upon arrival to replace the other one. If she were to arrive living, one would not expect that the Parisian climate, while it enriches those highly placed, would be favorable to African giraffes.

Fortunately for Zarafa, what followed would show the cynic to be wrong, while the journalist was certainly no dupe regarding the political dimensions of the gift: "Thanks be rendered to Muhammad Ali; he knows that small gifts maintain friendships; he sends us some quadrupeds instead of some salacious Greek ears." Rumor had it that in fact the *barbarous* Egyptian Turks were cutting off the ears of Greek rebels. Whatever the case may be, Zarafa encountered an irrefutable success: "The giraffe has been continuously followed by a crowd, since her release from the lazaret to the Prefecture where a thousand courtesans go each day, to see her arise, view her copious meals, her activities, her naps; it is the spectacle of the moment."

November 23, 1826: The prefect of the Bouches-du-Rhône sent the Minister of the Interior an account of the quarantine fees incurred in the lazaret. He would continue to do so each month until Zarafa's departure.

November 24, 1826: Anxious to conserve public funds, the prefect asked the customs director and the mayor of Marseille to exempt animals destined for the king's garden from customs and excise duties.

November 28, 1826: The professor-administrators of the Museum, Georges Cuvier, Mr. Cordier, and Mr. Bosc, thanked the prefect for care devoted to the giraffe.

It is worth noting that the aforementioned men misreported the event by writing that "over the past eighteen centuries, civilized Europe has not seen a living giraffe," although we know that Lorenzo de' Medici had pampered one in Florence in the sixteenth century. With foresight, they recommended that a "colored drawing be made that would allow us to know her exact proportions, and the contours of her head seen from the front and from the side: this would be, along with the skeleton and the hide, a small compensation for us if we had the misfortune of losing her before her arrival here."

They even drew up an estimated list of costs: "You will see that aside from the Black man, we thought that a second handler would be necessary, first to watch over her, then to replace him in the event he were sick, for we assume that the animal would prefer to be cared for by those it knows than by persons to whom it is not accustomed."

At the time, no one seemed to ask questions regarding the uprooting and availability of these Arab handlers. . . . But who were they? Bedouin in origin, the one called Hassan el Berberi was specially dispatched from the personal service of the consul of France in Egypt, where he held the functions of groom [*saïs*] and was promoted to giraffe specialist after his first similar mission to Constantinople. He was not allowed by Drovetti to travel beyond Marseille: "The groom Hassan would also like to make the trip to Paris with the giraffe, but since I am writing to ask you not to dispatch him to the capital during winter, it will be impossible for him to make the trip: I ask you therefore to send him back to me at the first opportunity."[7]

Hassan was seconded by the Sudanese man Atir, designated in the expense accounts as "the Negro, his helper," to distinguish him from the "Arab driver." The latter would become familiar with Paris. "If it is thought desirable to ask the Black Atir to care for the giraffe during her stay in Marseille and during her trip to Paris, and if it is deemed desirable to keep him on as handler of this animal, he is authorized to remain; in this case, he must cease to do so at my expense; I am also writing this to the Minister of the King's House, as well as to Mr. Bottu."

A third person, whose status was such that it was not deemed necessary to name him, escorted Zarafa across the sea; Drovetti designated him only as "one

CHAPTER 4

Figure 4.4. Visiting the giraffe.
© Aynaud Collection.

of my slaves." A fourth keeper was added in extremis: "As we had to embark three cows to nourish the giraffe, and seeing as how one of my people may suffer from sea sickness, I had to recognize the counsel of my friends to hire one additional man of service. This one is Egyptian and is [also] called Hassan."[8]

November 30, 1826: Middle-school students at the *Collège royal de Marseille* were authorized by the prefect to visit the giraffe. The three top divisions were accompanied by their deputy principal. Beyond this official request, it is probable that Zarafa's presence was a pretext for numerous high-society *soirées* (figure 4.4) that linked public relations with scientific curiosity. A woman from Marseille who had the honor of being present, said admiringly about Zarafa:

> ... her white skin decorated with reddish well-drawn spots, her elegant coat, her gracious shapes ..., her sweet, and even caressing demeanor. As I watched her, she fixed her gaze upon me, and I couldn't help calling her *ma belle*. With these words, she lowered her head which graced the ceiling and passed it before the low window where I stood. I believe that her intention was to lick me, for her tongue came out, but I stepped back, and she pulled her long neck away.[9]

During this period of settling in, an extraordinary event took place: Zarafa's attempted escape! The story was briefly reported by the famous journalist Méry, in the serial *Le Garde national*, on September 28, 1833:

The physiognomy of this corner of the Earth is more African than Greek; a short distance from the Spanish castle, you could imagine on a sunny August day that you were in the Sahara. A noble animal, the giraffe, made this error. That African exile, led through those burning desert sands, breathing the fiery air, thought herself in a native desert, and moving away from her guide, in a brusque movement, ran in large circles around this burning plain where she found once again her beloved Africa.

With wit, Alexandre Dumas would develop the anecdote in his *Impressions de voyage: le Midi de la France*, published in 1834:

When the giraffe landed in Marseille, she was sickly. Scholars declared that she suffered from sea sickness, but her driver shook his head and clearly explained, in Ethiopian, that what one understood to be sea sickness, was in fact homesickness. As the learned men didn't understand one word of her handler, they frowned, thought a moment, and responded that he could well be right. The Ethiopian, thinking they were in agreement, took his animal by the leash, and at noon, under a sunny thirty-five degrees C., walked along the sea and entered the gorges of Mont Redon.

As soon as the giraffe found herself in the midst of these barren, stripped rocks, she raised her head, opened her nostrils, stamped her feet, and suddenly seeing all around her a burning hot sand akin to her native sand, she thought she had returned to Darfur or Kordofan, and sprang, so wildly and joyously, that she pulled the rope from the hands of her driver, leapt over his head and disappeared behind a boulder.

The poor Ethiopian man ran off shame-faced to Marseille. This time, the scholars, seeing him all alone, understood that he had returned without the giraffe. From there it was not hard to jump to the probable conclusion that he had lost her: science ascertained this with all its ordinary certainty.

Two regiments were requested from the garrison commander; the two regiments surrounded Mont Redon, and found the giraffe lying stretched out on that beautiful African sand that had brought her back to life. The giraffe was too happy there to let herself be recaptured easily: but she was dealing with an able strategist. The colonel commanding the expedition was from Gémenos; he thus knew all of Mont Redon's passageways. After demonstrating feats of agility, the poor beast, surrounded by red pants, was forced to let herself be taken; she surrendered therefore with good grace to her Ethiopian, who brought her back to Marseille in triumph.

Never had she looked so well; one day in the sands of Mont Redon had sufficed to return her to health.

Doubtless embellished by Dumas's prose, and if the story is nonetheless confirmed as true, it would be logical to think that the prefect would take care never to officially mention such a misadventure that could have cost him his position.

CHAPTER 4

December 13, 1826: Two of Zarafa's guardians have already returned home, including Drovetti's nephew, as their services were no longer required. The official report made by Frédéric de Sillans, military departmental *sous-intendant*, stated this, indicating that "several of the effects that had been loaned for the use of these men were no longer needed, [hence] the necessity of returning them to the store." On January 5, 1827, the prefect, in a new list of expenses addressed to the Museum, added the "sum of two hundred francs, tallied by way of a gratuity to two of the four Arab drivers who accompanied the animals, and whom he believed on the advice of M. Drovetti, nephew to the Consul General, should not have to return to their country without this sign of the government's generosity." Neither history nor the French administration have retained their names, but they must have been Consul Drovetti's enslaved man and Hassan's helper.

In the same letter, Villeneuve also specified that "one of the three cows brought from Egypt was so exhausted that they had been required for some time to supplement the portion of milk she was providing, and it seemed better to buy another in order to reduce the daily expense, as cow's milk is rather expensive here."

And so the ailing bovine was sold, and the money from this sale was noted as a credit toward overall budgetary expenses.

5

THE ANIMAL UNDER A MAGNIFYING GLASS

An Aberration of Nature

The first images of Zarafa were scientific illustrations, requested by the museum professors, who were excited by this unexpectedly generous gift and also anxious to keep the specimen alive (figure 5.1).

The artist Philippe Matherou made the first lithography of Zarafa, taking care to provide a detailed text for it, in an account that stirred the imagination:

> The Giraffe, sent to His Majesty Charles X by the Pasha of Egypt, and in whose honor I offer the attached lithography to the public, hails from the area of Sennar, a large African city, Capital of the Kingdom of the same name in the Nubia, situated approximately three hundred fifty leagues SSE of Cairo. She arrived in Marseille in October 1826, on the prefecture grounds, from where she will be led to the King's zoos during the next warm season.
>
> Female, and about two years of age, she is the first living animal of this species to arrive in France. Her vertical height taken from the base of her hooves to the base of her horns or excrescences on the top of her head: ten feet, eight inches, that is, three meters, forty-seven centimeters. The length of her head: one foot, six inches, that is, fifty centimeters. Her body, her head and her horns are covered with short, fine hair. The coat has a white tawny background, a bit darker along her mane and hindquarters. The shade becomes lighter as one approaches the lower front areas, so that one sees the back part of the neck, the chest, the stomach and the lower legs as almost white, as are the inner thighs. She has irregular tawny patches, whose color becomes lighter as one approaches the aforementioned areas, and finishes about the same shade as the color behind them. The horns, the forehead protuberance and the face of the head, down to her upper lip, are a dark chestnut color.

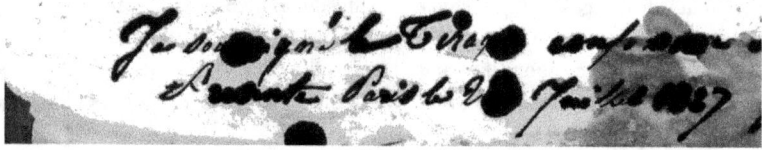

Figure 5.1. Giraffe in the spyglass.
© Musée Carnavalet—Histoire de Paris.

The surrounding area and lower mouth include a kind of sparse beard of straight hairs about 8–10 centimeters in length, of a yellowish grey color. The color of the lips is brown-black with different shades underneath and around the mouth; that of the ears is tawny white, the interior is a bit grayer. The side of the head is white with tawny spots; the underneath is a dirty white. The mane hair is supple and dark tan in color. The tail is a lighter tan with small, darker spots; it is covered with very short hair which ends in a ribbon of stiff, black hairs. Her hooves are a bluish, dark gray.

The Giraffe lives in Ethiopian deserts and several other provinces of Southern Africa. She kneels to rest, drink, and graze from the shorter grasses; she willingly grazes on buds and acacia leaves. She chews her cud. Her gait is an amble. Kicking with her forelegs is her main defense against aggressors. In fact, she is quite gentle in character, as her physiognomy suggests.

I must defer to the learned naturalists the task of giving a complete description of this animal, limiting myself to this cursory record, which serves as a guide to my drawing, I hope that these initial pencilings (my first sketches) will be agreeable to the public, and received with a kind indulgence.

Zarafa's iconography startles by its richness and diversity in terms of weight, color, style, and setting (figures 5.2, 5.3). A naturalist's description might have satisfied those whose goal was more zoological than artistic. But the subject matter was just too colossal, bizarre, and extravagant. The giraffe inspired artists and appealed to precision, or even perfectionism. The quasi-totality of the Zarafa illustrations of the time were enhanced by associated natural or decorative elements from the botanic, animal, or human realm. It is rare to find an image of the giraffe without her handler(s). Linked by the rope and by fate, Zarafa with her handler Hassan constituted a singular and complete concept. For today's observer, this association may appear degrading for the human in question; at the time in Europe, the duo passed as an inseparable paragon of African exoticism.

Mr. Salze and his fellow academics at the *Académie des Sciences* in Marseille examined the amiable monster from head to tail, including its excrement. Through an interpreter, they questioned the indigenous handlers at length, learning 1,001 details about the animal, and delivered the result of their investigations not only to the scientific world but also to the French people: "The appearance of a giraffe in France has offered those amateurs of natural history an interesting subject for observation, and a curious spectacle for the entire population. It is the first time in this country that we have seen such a gigantic animal species, so remarkable in its uniqueness, its curious contours, its behavior, its physiognomy."

Figures 5.2. Two drawings from the *Dictionnaire des Sciences naturelles*: one (left) before the naturalists saw the giraffe, and a revised version (right) after they saw it.

Figures 5.3. After seeing it, naturalists have revised their sketch.

In Cuvier, Frédéric, *Dictionnaire des Sciences Naturelles*, Planches, 2e partie, Paris, F. G. Levrault, 1818–1829. © Muséum d'Histoire naturelle de La Rochelle / Lézard Graphique.

CHAPTER 5

Zarafa's arrival, along with all that ensued, considerably enriched knowledge of this subject, which had been quite limited until then. An unknown creature had come to France, and the object of observation this time was not reserved to an erudite elite, holed up in their laboratory. Curiosity was not limited to the specialist, as information could be gleaned firsthand and was within reach of everyone:

> This animal's homeland is Central Africa. We don't know the extent of the lands it [the giraffe] occupies, and if we had to depend on Levaillant and Buffon, their territory would be very limited. The individual animal we have was taken not far from Sennar, in a mountainous area covered with thick forests, from whence the Nile appears. In this particular area, the presence of ostriches, gazelles, antelopes, panthers, and small lions has been noted, and a bit farther, elephants and rhinoceros. According to some Arabs, giraffes are small in number; they are seen most often in groups of three, two older and one younger.

Zarafa defied comparison with other known species, for she belonged to no other identifiable category:

> It is according to their shapes that science has divided natural creatures into classes and families, into genres and species, and the profound, hereditary modifications that living species have undergone have given science the means to establish races and tribes. This constant variety is numerous among domesticated animals, while rare among wild animals: the giraffe is among the very small number of those that evoke no other.

Upon closer examination, the giraffe in fact resembled neither deer, antelope, camel nor cow: "As to the ensemble of its anatomy, this confined body, positioned along such an oblique angle, with such an enormous neck, has characteristics which no other quadruped comes anywhere close to resembling." One observer tried to express the inexpressible:

> [The giraffe] really enjoys leaving her stable, and when she is led into the prefecture garden on nice days, which happens often, she jumps as if she were a filly, but in a manner so distinctive that it cannot be readily described. She rises up rather high, and then falls back down onto her stiff, immobile legs. Sometimes she wishes to gallop; she then drags along the four Arabs restraining her, and we have seen her drag five robust men in a moment of gaiety.

Her varied behaviors were fascinating:

> [They were] not less remarkable than her anatomy. She walks or gallops; one has never seen her trot: her step is a perfect amble, so that the body is constantly

in balance on one or the other of her two sides. Her stride is long, rather easy, but completely devoid of grace: her gallop is a series of jumps, and the animal is built so as to be incapable of sustaining this gait for long. Standing in her stable, the animal is never immobile like cows and horses, but is constantly rocking and rarely lies down, and akin to the manner of horses, she chews standing up; her sleep is extremely brief.

Even more fascinating than her unusual appearance was the gentleness of her manners, already praised by ancient travelers. However, while Zarafa adjusted quite easily to other animal species, this sentiment was not necessarily shared: "She appears to like horses very much but they do not reciprocate; and when she tries to approach in order to follow them, they paw at the ground, whinny out of fear, and try to run away quickly. This antipathy, this terror, are stronger still with mules. Cows who see a giraffe for the first time do not appear frightened in the least."

The giraffe showed a special capacity to enter into a benevolent relationship with people:

> Our giraffe has an extremely gentle disposition; she has still not shown any anger or malice. On occasion, she is a bit capricious; she is sometimes impatient, and that is when she drags or easily carries off three or four keepers by whom she normally allows herself to be led with the greatest docility. She walks around without appearing to perceive the flock of curious people surrounding her; she lets herself be approached very closely, provided she is not touched. She likes to lick the face, the hands, and the clothing of those caring for her, and it is a display of kindness she sometimes gives to strangers who approach her.

Humans can be hostile to the giraffe. We cannot ignore this point, as their capture in Africa has been described:

> They don't run off at the appearance of men; it is only when they are approached for capture that they flee, and with enough rapidity to leave behind even the best horses; but if they are pushed onto the plains, they are quickly seized, their chest does not have sufficient capacity, and they will run short of breath. Sometimes when out of breath, they do not surrender; rather, doing an about-face, they vigorously defend themselves with their forelegs. Desperate to master them, the Arabs kill them: they eat their meat, which they find to be exquisite, especially that of young giraffes. The Hottentots chase giraffes with poisoned arrows; they particularly enjoy the marrow found abundantly within the bones of these quadrupeds; they use their hide to make drinking pouches, which, despite the shortness of its hair, is an inch thick and very strong.

CHAPTER 5

Particular attention was paid to Zarafa's diet. We know already that nothing could disturb her milk-based regime. After twenty to twenty-five liters of daily milk, that is to say, six to seven gallons every twenty-four hours, she would spray a last mouthful into the air, signifying her satisfaction. She could also give lessons in hygiene, since her milk "is presented to her in a very clean vessel; the slightest odor could cause her to reject it. The Arab who served it to her also had to observe the strictest cleanliness for himself."

They were amused at the gymnastics required for the animal to drink:

> It was not without great difficulty that while standing, it could lower its mouth to the ground; it was obliged, in order to do this, to spread its front legs, retract its hindquarters, push its shoulders forward, lengthen its neck through a stiff, forced, and somewhat ridiculous move; one might call it dislocated, crippled: it can then pick up a fallen tree branch, but it is inconceivable that it could drink from this painful position. [And yet . . .]

At that time, it was notable that Zarafa refused to drink in the presence of strangers. And especially not water. Her disgust for it was explained by one of the handlers, if one believes the chronicler: "The repugnance of this animal for water was thus explained by this Arab: giraffes only drink from a large lake whose water is white, sweet, and slightly warm; they run to it from a distance, and to drink it, they wade into it up to their knees. In fact, one finds to the west of Sennar a great tributary of the Nile, called Baar el Abial: White River." It was in fact the White Nile, but Zarafa was to have come to Khartoum via the Blue Nile. Geoffroy suggested a more rational explanation:

> The Arabs, its first masters, imposed upon it the conditions that they themselves were necessarily subject to, or if you will, they were obliged to share their food, and the resources linked to their nomadic life. Thus, they first gave it milk from their camels, which they continued to do later on because, in the areas of the desert where they lived, it was easier for them to procure milk than water.[1]

After weaning, solid food was required. And so it was necessary to provide eighteen kilos of a daily mixture of corn and barley, with *mill-ground* broad beans. In addition to this, and during her strolls, the giraffe grazed with pleasure on the leaves of linden, cherry, mimosa, and black locust trees, delighted in yew and thuja, and couldn't resist tasting the ash trees, whereas she scorned the grass of the lawn: "As this animal is destined to live off the leaves of trees, Providence was not content with giving her a colossal size, but also provided her with the means to reach an object above her neck, by raising her head and stretching out her tongue." This tongue, measuring fifty centimeters, was so long, so black, and so agile that the first visitors imagined they saw a serpent emerging from the animal's throat.

They were mistaken about Zarafa's silence. "A final trait of this singular animal is an absolute mutism; no sound has yet been transmitted from her mouth, other than the light noise of mastication. Unless it has lost its voice in captivity, one could say that there exists a completely mute quadruped; but this would be the only one."

In the end, she was scrutinized, examined, and described in detail, but the giraffe remained a living enigma. Her body contained a mysterious equilibrium, a harmony and beauty, born of the marriage between incoherence and improbability:

> One can say that there is nothing elegant or gracious regarding the giraffe's particular contours: its short body, its high, closely grouped legs, the excessive length of its neck, the slant of its back, its poorly rounded rump, and its long, hairless tail, all these things contrast in a shocking way. She sits so poorly and is so unbalanced on her feet, yet we are seized with astonishment regarding her appearance, and find her beautiful without being able to say why.
>
> It is really remarkable that after having studied her attentively, we only retain an uncertain memory of her contours. This is why I believe we generally like to revisit her often, and each time, she evokes something new upon which to comment.

January 5, 1827: The prefect wrote to the Museum professors to update the accounts and began to prepare for the giraffe's journey to Paris.

In the capital on that same day, the Count of Peyronnet, in *Le Moniteur*, presented a text from the government for the reestablishment of censorship, disastrously announced as a "law of justice and love." Casimir Périer mocked it: "One might as well propose a unique article which stated that printing has been abolished in France to the benefit of Belgium." It stirred even the *Académie* to comment. Most of them softened certain aspects of the text, but the Chamber's debates were vehement. There were shouts at the ministers: "What have you done, until now, that raises you above your fellow citizens, so that you find yourselves in a position to impose your tyranny upon them?" In truth, at the time, only the giraffe was seen to rise above the citizens of France. And from her height, what she perceived regarding the future of their king was not rosy.

February 6, 1827: With each update of his accounts regarding the giraffe, the prefect made sure to inform the Museum regarding her health: "The state

CHAPTER 5

of the animals is still acceptable. The giraffe has maintained all of her liveliness. Winter, which was harsher than usual here, did not affect her in the slightest. The assiduous care provided to her has been a success, allowing one to hope for her preservation."

February 12, 1827: Bending over backward to provide regular monthly correspondence, the prefect dispatched a letter to relate the giraffe's weaning in progress: "The animal appears to be acquiring still greater energy: she shows signs of a lively impatience when the time comes for her usual promenade. It has even been said that following the example of the cows that please her, she is beginning to eat some hay."

On good days, a ritual began that all enjoyed, animals and people—aside from the constabulary, no doubt. At noon, during mild weather, Zarafa was led along the paths of the milk cows, for a daily walk of two or three hours in the surrounding countryside. She wore a collar with two double ropes so that Atir and Hassan could lead her.

> Since the return of good weather, few were the days she was not seen on one of her promenades, which successively led to all points of the area, where on occasion she caused the greatest surprise among the countryside's residents. All took pleasure in seeing her stretch to full length as she exercised the patience of her Arab drivers; she paused with each step to graze from the acacias on the boulevards, and they also liked to watch the movements of her head each time she used her very flexible neck to drop it within reach of a child to brush an extended hand, or to grasp the flowers or greens offered to her. Due to her popularity so rightly deserved, florists bedecked this beautiful African with pink broom flowers the moment she crossed their market on her way.[2]

March 3, 1827: The increasing number of curious onlookers was such that during her daily walks the prefect had to officially request an escort of two policemen from the chief of police. A singular, regrettable incident occurred, when one day, several horses hitched to carts took fright, injured a mule, and damaged two carriages. With finesse and foresight, the prefect compensated the victims:

> I thought it would not be right to refuse this indemnity, although strictly speaking one could regard the incident as a pure accident, inasmuch as every precaution had been taken and the drivers of the carts were forewarned; but as no other damage has been incurred here during the giraffe's lengthy stay, and unable to deprive her of daily strolls, in the interest of the animal's preservation, as they accustom her to taking long walks, it seemed to me necessary to prevent anyone from becoming indisposed toward her drivers.

March 4, 1827: The female antelope who traveled with Zarafa from Alexandria died. Two days later the veterinarian from the communal district of Marseille, Doctor Vial, was called by the prefect to perform an autopsy:

> I have found (1) a strong contusion of the right thigh, due to several head and horn butts given by the male antelope; (2) the omentum, the liver, and the lungs carry within them several hydatid worms from the thickness of an inch to that of a large walnut, and they can be attributed to the complete state of obesity in which the animal was found; (3) finally, there was partial inflammation inside the four stomachs, notably within the last one (abomasum). I do not know the cause of this gastritis, and I believe that all of these wounds together were more than sufficient to bring about death.

Ulcerous, worm-ridden, obese, and abused by its partner, the poor antelope had in fact good and multiple reasons for not surviving its emigration. It would therefore never see Paris.

The question of the day was henceforth the trip northward to the capital and delivery to the king, who started to become impatient. Opinions diverged as to the means of travel. The prefect was against a trip by sea, via Gibraltar to Le Havre, and then up the Seine River. Enamored with local history, Villeneuve undoubtedly knew of the misadventure of the rhinoceros of Pope Leo X.... In 1513 an Asian rhinoceros was offered to the king of Portugal, Manuel the Fortunate, by the king of Gujarat. Wishing in turn to offer the animal as a gift, the king of Portugal decided to give it to the pope. On the way to Rome in 1516, the convoy stopped in Marseille, on the island of If. At that time totally unknown to Europe, the rhinoceros piqued the curiosity of the island's inhabitants and that of Francis I of France, who came to admire it after his return from Marignan.[3] Unfortunately, the pope would receive a stuffed rhinoceros, as a violent storm crashed the ship upon the reefs of the Gulf of Genoa soon after leaving Marseille, and the animal's cadaver washed up on shore.

The prefect also didn't foresee a trip across land. He admitted this to the professors on January 5, 1827:

> Even though there is still time here before the temperature will allow for transport of the animal, I think that we might already consider choosing from the available options. The trip by sea would be too dangerous, for there is reason to wonder how the trip from Alexandria to Marseille was accomplished with such good fortune. I also don't believe that the Giraffe could make the trip across land from here to Paris: aside from the obstacles that her appearance and habits would create, too many accidents and difficulties could ensue from her encounters with

CHAPTER 5

carriages and the large quantity of onlookers. You will perhaps judge her embarcation on the Rhône river and the continuation of the trip by internal navigation as most appropriate.

March 9, 1827: In his letter posted from Alexandria, Consul Drovetti declared himself against the choice of roads and against excessive usage of riverways: "The trip across land from Marseille to Paris could present dangerous occasions for the Giraffe, as well as those responsible for leading her, I think it would be preferable to embark her for Le Havre, in order to avoid at the same time the inconveniences that would be created by seeking to navigate the Rhône."

March 15, 1827: The Museum professors drew a blank. . . . They wondered if it would not be preferable to designate a competent person to make the best decision and manage the operation. They then suggested to the prefect a sire too zealous to be trusted:

> The moment has arrived when we must undertake transportation of the Giraffe. The importance of such an acquisition for Science and for the Museum Zoo, the measures taken on her behalf by the Minister of the Interior, and especially the interest that we know the King takes personally in the preservation of the precious animal offered to him by the Pasha of Egypt; all this makes it our duty to take the utmost care in the choice of means to utilize in order to ensure the success of this difficult operation.
>
> You have yourself, My Lord Count, showed in this circumstance too much deference towards the Museum, and we are already too indebted to your informed care for us to hesitate a request for this again at present.
>
> We ask you therefore to please let us know your opinion regarding the best means to undertake this transportation, either by land or by the Rhône, and all the preparations necessary for this trip. Perhaps it would be advantageous to send along an intelligent and able person to direct the operation.
>
> The owner of an ambulatory zoo, Sire Polito, who is currently in Marseille where he has seen the Giraffe, has proposed to us to shoulder responsibility for transporting the animal by land, noting that travel by water, as he said, would create numerous drawbacks, principally due to changing boats.
>
> This proposal, due to the experience he must have acquired with this type of travel, appeared to us to merit further examination, and we would be grateful if you would be so kind as to ask Sire Polito for some detailed explanation as to the means he would intend to use and the fees he would require, in the event that we believe he should be put in charge of this undertaking.
>
> It would be worrisome if he had calculated principally on the profit he would make in showing the animal during the journey, which would in no way be suitable for an item belonging so directly to the King [figure 5.4].

Figure 5.4. The giraffe's promenade.
Former Collection Henri Georges, Public Domain.

Upon reading the letter, the prefect quickly noted in sharp terms in the margins, "as for Sire Polito, it is a matter of speculation on his part, and there is no reason to follow up on his request."

March 19, 1827: Four days later, Villeneuve wrote to the professors, though he still had no knowledge of their letter dated March 15—at that time, postal *diligence* had not really earned the full measure of its name. He began, as usual, with an accounting of current expenses and then noted the loss of the antelope. Based on the veterinarian's assessment, he defended himself against any potential suggestion of lack of care and affirmed his application of the procedure required in such a serious event: "I have recommended preservation of the animal's hide and skeleton."

Then he reassured them as to the state of the Giraffe:

> [She] is still full of vigor, and since it is no longer cold, we lead her some distance away from the city each day. We have recognized that she required this exercise. We are reassured by these walks, for which six men are responsible for holding her as a precaution, by means of as many leads, that this animal easily lets herself be led. Her companions, the cows, walk before her, and she obeys their direction. Neither noise, nor carts, nor the crowd of curious onlookers that gathers around her cause her the least distress. The animals that encounter her are similarly without fear.

CHAPTER 5

But here lies the point of the prefect's letter:

> All these observations lead me to envisage the possibility that she could be led to Paris via short, daily walks. No other means appears preferable to me. Traveling by river, which I had initially considered, would present disadvantages linked with many difficulties, for in passing from one surface to another, she would run the risk of a fracture.

The prefect was too urbane and too much the public servant to impose a solution, yet he had outlined an impeccable argument. So in conclusion, he put everything in the hands of the Parisian specialists but agreed with the idea of these great minds that they send him "a person from Paris whose comparable knowledge and aptitude would inspire total confidence, and to whom you would confer responsibility for the care of watching over the transfer of the Giraffe, a suitable person who would come and study some of her habits before departure." Finally, he concluded in assuring them that he "was making haste to enable the trip to begin before April 25." This estimate proved to be overly optimistic.

March 26, 1827: The prefect took up his feather pen to respond to the letter that arrived before his did. He confirmed and detailed his plan for the route: "I will make sure to alert the Prefects of the Districts to be crossed regarding the passage, and to request a police escort for the route."

He then reiterated the urgent need for a specialist to be dispatched from Paris and made his opinion clear regarding Sire Polito: "That owner of an ambulatory zoo is no better in fact than any other for the precautions to be taken to preserve an animal generally unknown and which requires a particular treatment that must be studied. You have correctly presumed that this proposal was made with the intention of profiting from public curiosity."

April 5, 1827: The Museum professors did not take sides: "The observations you have well thought to send us regarding the Giraffe's travel have led us to think as you have that it would be difficult to make a definitive decision on this matter, without consistent study of the animal and its habits."

Nevertheless, they did name an authorized person to undertake this mission:

> We have thus made the decision to assign responsibility for this examination and ensuing determination to the Museum's Professor of Zoology, the one among us whose qualities for both are a natural fit. The Professor, Mr. Geoffroy Saint-Hilaire, has consented to take responsibility for this mission, and he intends to depart for Marseille in the next few days. If, however, some unforeseen circumstance were to retain him in Paris, Mr. G. Cuvier, Keeper of the Museum's Zoo, would be charged with replacing him in this instance.

6

BETWEEN SCIENCE AND RELIGION

An Animal at a Price

At fifty-five years of age, the famous French naturalist Étienne Geoffroy Saint-Hilaire (figure 6.1), director of the Paris zoo, professor of zoology and member of the *Académie des Sciences*, agreed to return to the field by assuming responsibility for the first giraffe in France. Who else could better accomplish this royal mission? Thirty years earlier, he had followed Napoleon on his Egyptian campaign and was certainly the only scientist in France in 1827 to have gotten close to a live giraffe.

Figure 6.1. Étienne Geoffroy Saint-Hilaire, ready for his shot at glory.
© Musée intercommunal of Étampes.

CHAPTER 6

For Geoffroy Saint-Hilaire, the study of the giraffe would be a living demonstration of his celebrated theory of "the balancing motion of organs": "A system of organs does not acquire a dimension beyond communal proportions, unless there is the necessary reason to do so when other organs are restricted and diminished by an equivalent amount."[1] This was why the giraffe possessed a short head and trunk but long legs and neck. "She uses her elevated head as a kind of balancer, to carry an excess of weight to one side as required."

At that time, the French scientific world was in heated debate over the origin of species. Lamarck revealed his conviction in 1809 in his *Philosophie zoologique* that the evolution of the species was a recognized fact. His theory of *transformism* centers on the influence the environment has on organ development and their modification by type of usage, and also on the heredity of acquired characteristics. Zarafa would be of great interest to Lamarck, since he had referred to the giraffe to support his thesis even before her arrival in Paris. He stated that the lengthening of the giraffe's neck was due to sustained use:

> Regarding habits, it is curious to observe the giraffe (*camelopardalis*) with its particular shape and size: we know that this animal, the biggest of all the mammals, lives in the African interior and in places where the ground, almost always arid and without grasses, requires it to graze on foliage from the trees, and push itself up constantly to reach for it. The result of that habit, maintained over a long period, in all individuals of its race, is that its forelegs have become longer than those behind, and that its neck has so elongated that the giraffe, without rising up on its hind legs, can raise its head and attain six meters in height (almost twenty feet).[2]

Further developing his theory, Lamarck noted that his explanation was logical for yet another reason. We know that the earth has changed continually over its long existence. However, since one species, in order to survive, has to be in harmony with its environment and since this environment is constantly changing, species must also evolve in order to maintain balance. Lamarck thus discovered the flaw in natural theology and *creationism*. It was possible to imagine that God had created living organisms perfectly adapted to their environment, but only in a static world. How could some species remain perfectly adapted to an environment continually in flux? This was possible only if they transformed themselves by adapting to their milieu. QED.

In contrast with the Count de Buffon, a renowned eighteenth-century French naturalist who tempered his own positions, Lamarck did not fear intellectual critique. He would suffer its harsh consequences. His novel ideas underwent numerous attacks from the devout and the *fixists*, who allied themselves against the naturalist. Lamarck came up notably against Georges Cuvier. In this scientific quarrel, Geoffroy took Lamarck's side against his colleague from the Museum.

Geoffroy Saint-Hilaire, the founder of teratology, the scientific study of monsters in nature, upheld that the organization of animals was subject to a general plan whose detailed modifications resulted in their species. Every species could be materially deduced from a relative species, and an undeniable relationship existed between them. For Cuvier, each species was the fixed and unchanging product of a special creation, without possible passage between them. For him, all genetic modifications were an act of God, and they occurred during large natural events such as cataclysms, which were the work of the divine.

The stakes and dangers of a debate calling into question theological underpinnings are obvious. The government of Charles X was clearly reactionary. The secretary of state, *Monseigneur* Frayssinous, had a large number of professors dismissed. Geoffroy's scientific stature was too important to give him cause for personal concern, but he must have worried for his students' sake, and first and foremost, for his son and disciple, Isidore.

Taking charge of the giraffe's delivery to Paris would be Geoffroy Saint-Hilaire's last mission! Crippled by rheumatism, the elder scientist regained his youthful ardor to organize and lead this expedition. He set forth on the same road that his former emperor had taken eleven years earlier, in an attempt to regain power. Geoffroy, creator of the first French zoo, would innovate by leading one of the most fabulous animals ever seen in France over hills, and across valleys, through cities and villages. He would bring science, exoticism, and the marvelous to the common people. His greatest class would be taught outside the classroom; it would be his best lesson, the one that would mark his entire life.

April 27, 1827: Far from Parisian disputes, but just as passionate about naturalist novelties, the prefect of the Bouches-du-Rhône accepted the gift of two mouflons (wild sheep, *muffoli*) for the royal zoo; they were offered by his cousin, the Count Félix d'Albertas. These animals would accompany the giraffe's procession: "I am entrusting a groom to go and bring these two animals back from Gémenos."

April 28, 1827: The prefect wrote to the Museum, expressing some concern regarding Geoffroy's delayed arrival, as his departure had been announced three weeks before:

> However, it is worth noting that we have arrived at the most favorable weather to make the trip. As the animal should only travel for a short time each day, it will take a considerable length of time before she has made it through these southern lands,

CHAPTER 6

and for the warmer season to be well established in the north. I think it would be also good to avoid the very strong heat which arrives here early.

The state of the Giraffe is still excellent. The current temperature and green fields encourage a most amiable liveliness seen during her daily promenades. The continued care she receives makes it seem promising to us that you will be able to ensure the Museum of such a precious acquisition.

April 29, 1827: The government continued its measures of retaliation against opposition, which only served to galvanize it. Guilty of having demonstrated against the current power, the Parisian National Guard, created in 1789, was dissolved.

May 4, 1827: Geoffroy finally arrived in Marseille. One of the reasons for his delay was the naturalist's stay in Montpellier to negotiate acquisition of a collection of fish "brought back fifty years earlier by scholars who had explored the South seas in the company of Captain Cook."

The prefect immediately sent a certain Barthélemy to his cousin, Count Félix d'Albertas in Gémenos, to take delivery of the mouflons offered to the zoo, "which I will immediately put into the care of Mr. Geoffroy Saint-Hilaire, professor at the Museum, arriving from Paris to take charge of the Giraffe's delivery."

May 17, 1827: The prefect sent the Museum the expense reports for April and noted, "Mr. Geoffroy Saint-Hilaire has been in Marseille for several days, and we are making preparations under his direction for the animal's upcoming departure."

The same day, he wrote to the mayors of municipalities within his department:

A delivery of various animals, including a Giraffe, is going to be sent from Marseille to the King's zoo garden. The mission given to Mr. Geoffroy Saint-Hilaire, Museum professor, to personally oversee their transportation, is sufficient indication of the interest the Government attaches to the preservation of these precious items of natural history. I ask that you provide, either for the journey or the stay in your municipality, all aid and assistance the circumstances may warrant.

Things then developed quickly. Geoffroy conveyed his plans for the journey north:

Plan for transportation of the giraffe and a new pair of ruminants sent from Egypt with her:
  Needs

    –cart fitted with horse for the antelope's transport
    –cage to keep this animal
    –canvas cloth oiled or waxed for its [giraffe's] body and neck, in two pieces

–food supplies as far as Lyons, consisting of mill-ground broad beans, barley and Turkish wheat.

Personnel

–the two Egyptians and Barthélemy, ongoing caretakers
–per diem for three additional men requested from each main stop, to drive the cart and hold the Giraffe's leashes
–two military police officers, to keep order

(wages to be paid in advance either per day, or according to the distance traveled, for food and work).

Route to take to Lyons: Aix–Avignon–Orange–La Palud–Montélimar–Loriol–Valence–Vienne–Lyon.

Avignon is eighteen and a half leagues from Aix; where to spend the first and second nights, in order to ensure the greatest progress made by the animal? . . .

Outline of the giraffe's projected journey. Leaving from Marseille, the convoy will arrive in . . .

1 day in Aix	(8 leagues)	1 day
3 days in Avignon	(18½ leagues)	3 days
1 day in Orange	(7 leagues)	1 day
1 day in La Palud	(7 leagues)	1 day
1 day in Montélimar	(7 leagues)	1 day
1 day in Loriol	(6 leagues)	1 day
1 day in Valence	(8 leagues)	1 day
3 days in Vienne	(18½ leagues)	3 days
1 day in Lyon	(7 leagues)	1 day
	87 leagues	13 days

There would be a stay, as much for the giraffe's rest as for leaving to discretion public curiosity along the way, as follows:

in Aix, 1 day	1 day
in Avignon, 2 days	2 days
in Montélimar, 1 day	1 day
in Valence, 1 day	1 day
in Vienne, 2 days	2 days
in Lyon, 3 days	3 days
Walking	13 days
Resting	10 days

23 days of travel to arrive in Lyon.

The naturalist's forecast was twenty-three days from Marseille to Lyon and twenty-nine days from Lyon to Paris, for a total duration of fifty-two days. They would proceed by short trips to give Zarafa the impression that it was not so much a journey but rather a series of recreational promenades. It was also

CHAPTER 6

the most popular and least costly solution. As a pragmatic scientist, Geoffroy thought to have a raincoat made for the beautiful orphan: for 103.50 francs, a tailor sewed a custom-made two-piece suit from oiled canvas, buttoned in front, with a hood for her head.

May 18, 1827: A driver and his cart were commissioned:

> The undersigned Jean Chapsal is hired to transport an antelope and two mufflons on his cart from Marseille to Paris, in two cages to enclose these animals, as well as grain for their food during the trip and the driver's four bags.
>
> He will conform to the days of travel, rest stops and other circumstances of the journey, as per orders given to him by Mr. Geoffroy Saint-Hilaire, Museum administrator, commissioned for the management of this transport.
>
> The present commitment is undertaken for a price of seven hundred francs, that the Museum Administrator will pay him for all fees of any kind as soon as the transported items are delivered to the destination site of the King's Garden, after a period of time, as required for the journey and stay, nonetheless with the condition that if the period should exceed fifty days, he will be paid for this transportation, in addition to the agreed upon sum mentioned above, that of fourteen francs for each additional day.

The same day, Villeneuve wrote to the subprefect in Aix: "I ask you, please, to ensure by all possible means that every measure will be taken to preserve these items of great interest while on the roads. It would be appropriate for you to organize this with the Mayor of Aix to whom I have just written."

Then, to the department's head of police:

> The assiduous care that the police force has taken with regard to the giraffe's health during her daily walks are a gauge of what we may expect until her departure from this department. I ask you to take measures so that an adequate escort will watch over these precious objects of natural history, led by Professor Geoffroy Saint-Hilaire, to forestall along the road and during their stay any accident that might occur with regard to carriages encountering the Giraffe.

During all these efforts in preparation for departure, Geoffroy found the time to study not only the giraffe but also her companion, the antelope. In addition, he agreed to meet doctors and scientists in Marseille, and then in Toulon, those in residence there or students, avid to learn about the latest scientific advances in their discipline. Several collectors opened their cabinet of curiosities to ask for names to place on their possessions. As a dedicated pedagogue, the famous naturalist fulfilled these requests with a loquacious enthusiasm.

Once Drovetti's nephew had left, Geoffroy lacked an interpreter to effectively communicate with the Egyptian handlers. When he told the prefect about this, the latter had a brilliant idea. On the outskirts of Marseille, there was a camp of Egyptian refugees, including the Mamluks, survivors of the imperial guard and their families. These Mamluks, indigenous allies who had fought alongside Napoleon during his Egyptian campaign, became henceforth personae non gratae in their native land. History has demonstrated on many occasions that invaders have difficulty showing gratitude to their indigenous allies. Placed under police surveillance, the Arabs in Marseille were not well regarded and were even mistreated, by a population that was essentially royalist. Although they had returned to civilian life, they depended upon the military administration.

The prefect found among former Christian soldiers of the Egyptian Legion, "one Egyptian of good conduct," the ideal interpreter for Geoffroy. Born into the first generation of émigrés, the fortunate chosen one was called Joseph—a name that implied integration—Ebed, a common last name in his homeland. He was assumed under the "regulation of the deposit of Egyptian refugees, four hundred eighty-seven in number, being supported with daily aid amounting to seventy-two centimes each." The scholar would continue to call him by his Arab name, Youssef. On the 30th of May, the prefect intervened so that the indemnity would continue to be paid to Joseph's father, in the absence of his son:

> This young man, very useful here as interpreter for the Arab drivers of the Giraffe offered to the King by the Egyptian Pasha, has just been recruited by Mr. Geoffroy Saint-Hilaire, Museum administrator, to continue this work all the way to the royal zoo in Paris. He has been promised that at his departure one would solicit in favor of continuing his aid benefit, so that his father might receive it on his behalf. Please do whatever is in your power so that this promise is upheld.

After several administrative digressions, on June 9th, the lieutenant general agreed to the prefect's request, specifying that Ebed senior would receive his son's indemnity, provided that he could "produce a Certificate of Life," proof that he was still living.

May 19, 1827: Everything seemed ready for the big departure planned for the next day. Geoffroy Saint-Hilaire wrote to his host, the prefect and count Villeneuve. His expression was less coherent and logical than that of the prefect; the writing was messy. We can imagine this man, overwhelmed by his responsibilities and his emotions, anxious to satisfy those who had hired him. He used no capital letters, including for city names, and used many semicolons. The writing

CHAPTER 6

was awkward and the spelling sometimes approximate, but on the eve of this crucial day, one could surmise a certain excitement on his part:

> I have been so very pleased by your infinitely gracious reception, by the most attentive hosting, not only by you, but by Madam the Countess and everyone close to you. I am filled by this at my departure; I will not search for words to express the depth of my feelings: may I one day be called upon to clearly express to you, to demonstrate at least some measure of such a profound sentiment and so rightly evoked.
>
> I carry with me your precious work, *Statistique du département des Bouches-du-Rhône*, I here and now acknowledge receipt of it; the administration of the King's Garden should thank you and will be informed as to the favor that you have done for them, their thankful regards will be as swift as they are deserved.
>
> As Prefect, you can appreciate all the care I have taken in Marseille to devote myself entirely to the tasks with which I am charged: I have not been diverted by any pleasure or distraction; and while the giraffe has not required direct care on my part, I am a naturalist and have connected with all the naturalists and scholarly doctors in your city.
>
> As for the Giraffe, to whom I am principally devoted, I have followed your valuable instructions: I arrived in Marseille on the 4th of this month, I took the three following days to study the behavior and habits of this gigantic animal, in order to fully understand her sensitivity regarding means of transport: to this end, I have followed her during promenades and noted her demeanor on the road as well as in busy parts of the city. This is how the means of transportation could be decided with some assurance to the success of the undertaking.
>
> She can travel by foot and can leave right away, which I will certainly have her do. But this great personage has an entourage, milk cows as well as a very interesting companion. My main thought was, if it had been one of a number of antelopes often sent to and observed in Paris, one might leave it in Marseille and lighten the giraffe's load. It would also constitute baggage complicated by the transportation of food for this animal [Giraffe] which ingests, besides her supply of milk, various grains which have to be first mill-ground. There would be no hope of finding this at every stop along the way, or other foodstuffs that would be well suited to the sensitivity of this animal's habits.
>
> But the Giraffe's companion turns out to be a very precious animal: it is novel with regard to Zoology, and moreover, by its physical shape and structure, one would say it is a new sub-genre; for it brings together the main traits of the gnu, the mouflon and the antelope. In this circumstance, the Sennari (we will give it this name) requires some equipment. You must know that it is stocky and dangerous, it will require a solid cage; this and the giraffe's supplies have rendered necessary recourse to a cart. As some space still remained on this load, my Lord, you agreed

to take the offerings of the Marquis d'Albertas, to give the king a male and a female mouflon, the latter pregnant.

However, we have had to organize the supply of carts and cages for the animals, and from the 7th of this month until now, the workers have been continuously engaged, following the orders you have provided.

Due to these circumstances, my Lord, I have not been able to depart until now. All will not be equipped before tonight, and I will leave tomorrow, the 20th, at the break of day. Despite your repeated orders and my exhortations, the workers could not have accomplished this any sooner.

On this point, my Lord, you have complete knowledge. It is likely not known in Paris and one must be surprised to learn that the Giraffe is not already on her way. On appearance, I should certainly be accused of sluggishness, or who knows, perhaps even negligence. My Lord, as your elevated social position places you in a high position of oversight for the government, you will perhaps wish to come to my aid and assure his excellency the Minister of the Interior, that I have acted with as much diligence as I could.

Perhaps it would be of interest to note for his excellency that I have been employed in an entirely consistent and tireless manner, and in every instance in which service to the Giraffe left me available to do so, I visited all private and public collections of natural history in Marseille (and Toulon from the 8th to the 10th of this month); I was asked to provide names for the objects they possessed whose scientific denominations they did not know. I did this continually in both cities. Hospital doctors, from the navy hospital in Toulon to the civil hospices in Marseille, wanted me to bring to them current information on new research in anatomical science; instructors and students wanted explanations that I could not refuse. Several times then, two-hour improvised lectures in clinical conference halls or hospitals seemed an appropriate response to requests made of me. I found myself thus able to unite southern scientific communities to urban establishments. I think this is enough without pursuing this further.

My Lord, I ask you to receive the homage of my profound gratitude and my most perfect and respectful esteem.

May 20, 1827: The departure from Marseille took place under a rainy dawn, which offered an occasion for the giraffe to don the oiled canvas for the first time. The procession included the following, in order: a dispatch of two mounted military policemen, preceding Zarafa by five hundred meters, to clear the road and stop stagecoaches, postal carriages, merchant and peasant carts; the corporal and three mounted policemen (the relay escort from each county); three milk cows, led by Joseph/Youssef; Geoffroy (when he walked); Zarafa, surrounded by the head groom Hassan in front, Atir on her right side,

CHAPTER 6

Figure 6.2. The painter Brascassat immortalized the passage of the giraffe in Arnay-le-Duc. This oil on canvas painting is the only visual evidence of the procession.
Jacques Raymond Brascassat, *Le passage de la girafe à Arnay-le-Duc*, 1827, Inv. 887.2.1. © Musée des Beaux-Arts of Beaune. Muzard Beaune Photo Workshop.

and Barthélemy Chouquet from Marseille on the left side; a carriage, purchased at Geoffroy's request; a cart with baggage, supplies, antelope, and mouflons.

Confusion among modern chroniclers led them to write that the supply cart also served as a vehicle for Geoffroy Saint-Hilaire when he wished to rest, but the expense account separates the rental of a cart and its driver from the specific purchase of a carriage–doubtless the one seen at the head of the procession on Brascassat's painting (figure 6.2). There was no need to add anything to the already picturesque sight.

On the same day, the convoy arrived in Aix, where it sparked enthusiasm: "She was led through the main streets; they applauded when she deigned to graze from the exposed, second floor window flowerpots. She was lodged at the Mule-Blanche Inn, whose entryway door was inordinately high."[3]

Le Cérémonial d'Aix preserves in its collections the *Relation de l'arrivée à Aix d'une girafe, donnée à sa Majesté Charles X par le Vice-Roi d'Egypte* (Description of the arrival in Aix, of a giraffe given to his Majesty Charles X by the Viceroy of Egypt):

BETWEEN SCIENCE AND RELIGION

The mayor of Aix, Mr. d'Estienne du Bourquet, having asked Mr. Geoffroy de Saint-Hilaire, Professor at the Museum in Paris, commissioned by the Government to watch over and preserve the giraffe, to allow residents to encounter such a rare item, orders were quickly given to allow the public to enjoy her presence.

At the very moment she was urged to leave the shed at the *Bras d'Or* where she was sheltered from exclamations, the immense crowd, excited by the sight of an animal that dominated over them due to its gigantic stature, unleashed cries of surprise from all sides, echoed by groups at the windows and from the rooftops of houses overlooking the Rotunda. Four black men, one of which assigned to his Majesty was very handsome, were guiding the giraffe as she circled around the Rotunda, then the Courtyard, where she stopped to graze on acacia flowers. She was led onto the *Place des Carmélites* and the *Rue de l'Opéra*, location of the office of the First President of the Royal Court where she would enter, onto the *Rue des Jardins*, along the grand boulevard, then finally onto the *Rue de Lacépède*, before the mayoral residence. Similar cries of surprise and admiration were expressed along her way and during her promenade in the city, as outside the city; she was accompanied by two cows whose milk was her only drink.

At 7 p.m. she was led out for another walk. The public showed the same haste to see her again. Finally, the following day at 8 a.m., the convoy began its walk to Lambesc, escorted by several mounted police.

For her departure, they sang as she passed a lament from which a few couplets have been retained:

La girafe, on vient de partir,
Consolez-vous, jeunes fillettes:
La girafe on vient de partir,
Les Égyptiens l'ont conduite à Paris
Cet animal charmant
Lève le cul, baisse la tête
Cet animal charmant
Lève le cul, montre les dents
Il reviendra peut-être,
Cet animal plein de bonnes manières
Il reviendra peut-être,
En retournant au grand Sahara.

[The giraffe has just departed,
Console yourselves, young girls:
The giraffe has just departed,
The Egyptians have led her to Paris

CHAPTER 6

That charming animal
Lifts its rump, lowers its head
That charming animal
Lifts its rump, shows its teeth
Perhaps it shall return,
This animal with such good manners
Perhaps it shall return,
By returning to the great Sahara.]⁴

May 21, 1827: In his first letter to the prefect since their departure from Marseille, Geoffroy constantly felt the need to justify himself, as if he feared being sanctioned by a superior:

> Our beginning was difficult; with a great distance to walk, an enduring rainfall, a crew not yet adjusted to the needs of the trip and rather disposed to liberty; these initial elements have in no way dismayed the giraffe, have not at all disturbed her health and are to the contrary encouraging for the future, where good habits will be required and I hope will be adopted. All that proves, as the saying goes, that there is a God for the innocents. People could not believe the arrival of the Giraffe in Aix: they said everywhere, by whose extravagance has this animal been put on a trip in such inclement weather? Madam Countess, whose goodness is so perfect and apt foresight so well demonstrated, should come and tug upon the ears of that extravagant one to enact a great and good justice! This is my entire defense, my Lord, that I was in good standing to have merited help from the god protector of innocents. I don't insist; I thus surrender my weapons. In truth, I had wished to depart the very moment it was possible to leave. In Paris, I am surely thought to be negligent, because only the goal is sought: the giraffe is desired, and one is hardly disposed to think of the difficulties.
>
> My Lord Count, how I am personally indebted to you! May none of this goodness ever be forgotten! I write to you today, regarding the beautiful African, which is rendered unnecessary by the return of Mr. Bazin, who has been a perfect traveling companion, and who is returning to you, only in order to renew the promise of keeping you informed of the compliments that this female will receive on her journey, due to your affection and attentions.

May 23, 1827: The subprefect in Aix-en-Provence gave an account of the giraffe's passage in Aix, to his superior. the prefect and count Villeneuve:

> The convoy of animals, destined for the King's Garden, left yesterday morning at 8 a.m.; the Giraffe was in perfect health; Mr. Geoffroy has been very pleasant with

the crowd; on Monday morning, he had the Giraffe walk around, and she was exhibited for people to view for some time: the same walk took place in the evening at 7 p.m.; the line of curious onlookers was unbelievable.

In accordance with your instructions, Mr. Geoffroy had the French coat of arms added to the waxed canvas coat used by the animal to cover herself during the trip.

Crossing French territory, Zarafa henceforth became French; Villeneuve decided that it would be good to display this identification.

The same day, continuing to manage the event at a distance, Villeneuve wrote to the prefects of the departments of the Vaucluse, Drôme, and Isère:

> From the 20th of this month, the foreign animals that spent the winter in Marseille have been on the road to the royal zoo. One of them is of such great interest that Mr. Geoffroy Saint-Hilaire, Professor of Zoology, an administrator of the Museum of Natural History, has been charged with watching over it during the journey. I am talking about the Giraffe. This precious animal has received continual care here over the past six months which has achieved the best results. Continuing this care, I believe it necessary to advise you regarding her upcoming passage through your department. The Giraffe is very gentle and will not provoke any accidents other than the occasional shock at seeing her gigantic body, or outbursts from horse and carriage that may be frightened by her appearance. These two points indicate the nature of the precautions that concern the authority responsible for public health measures. Three military policemen taken successively from brigade to brigade have seemed to me necessary for this department. If you think it suitable to give instructions to the mayors along the way, they should principally assume responsibility for designating a stable whose plank [ceiling] is at least twelve-to-thirteen feet high, and to place several cows nearby for the drivers. The Giraffe drinks nothing but the milk from these animals and it is given to her as soon as they are milked.
>
> Your keen understanding will dispense me from providing many details; other requirements may vary according to the situation.

The journey continued. In Saint-Cannat, a reception was given for the arrival of the "Princess from Egypt." Then in Lambesc and in Orgon. The countryside changed: pines and rocks took the place of cherry groves, almond and olive trees. The crowds did not change, growing in proportion to the cities crossed.

May 24, 1827: Geoffroy Saint-Hilaire wrote to the prefect from "Désiré, situated between Saint-Andiol and the Durance River." The progression was moving well: the scholar was in a good frame of mind, which allowed him to engagingly narrate each of the inevitable hazards with humor:

CHAPTER 6

So we have reached the border of the Department that you govern; after our two-hour break, we will cross the Durance and arrive in Avignon around 4 p.m.

It is difficult to advance more quickly: everyone knows what he has to do, and each one is busy working; I say this with regard to both animals and people. This morning in Orgon, the giraffe was calm under her emblazoned cape: as soon as she saw the cows departing, she began to stir, preceded by the order of her head groom Khassan [Hassan], glorious as a peacock, holding the leash attached to her head. He feared that Barthélemy from Marseille would presume to take the lead role, i.e., in that Arab man's thinking, to act as his superior while the black groom Atir was entrusted with second place position, holding the leash on the right side. Barthélemy is the left side keeper, and the smaller black man Youssef leads the cows. This order was followed quite precisely, each of the animals and people having well understood the spirit of the traveling conditions.

This has resulted in perfectly balanced health for all; the military police have been careful, very punctual and helpful in every way. Within the Department, there is a high intelligence that doesn't flaunt its actions but discreetly orders everything efficiently. It is no accident, just as it was no accident that all things here have been arranged, and so harmoniously that one must consider the cause of so many wonderful results.

I have insistently asked the subprefect of Aix to write you about the explosion of sentiments the first viewing of the Giraffe produced on Monday, among the whole population gathered along the way. We had promised that the animal would be shown at around 10 a.m. We had arrived well before this, but we had to wait for the passing of the procession for rogations. We didn't know the reason: the crowd's impatience was provoked. It was then even better prepared for the effect of the display. As soon as the Giraffe appeared in the midst of an immense gathering, there was one cry shared by all, but it was prolonged and especially deafening. They gazed with admiration upon this animal that dominated all human forms, confident of its value above this crowd of people, majestically swaying its head in the air and grazing from the treetops. But the population was insatiable, and the giraffe was more tired from its activities during rest periods than from the day's walk. I noticed that giving her over to the public's gaze meant that those of the working class, strong in temperament, mood, and strength for disputes, kept their spots up front. So we had to find other means to facilitate viewing by the more discreet and well-brought-up Bourgeoisie, resulting in double the work for the poor little girl your administration has adopted.

Presently, we have had to defend ourselves from a few swindlers along the way; they sank their claws into the Giraffe;[5] we are paying for our coat of arms: the king's beautiful animal, one says, has a master well able to pay; some went as far as to dub me the Count of Saint-Hilaire, while another called me Count of the Giraffe, treating me less royally.

One doctor could not recognize me in my current guise and moreover, I almost lost one monster he had been saving for me. As for the collection, my harvest has been abundant. Two [monsters] in Aix, one in Saint-Cannat, two in Lambesc: I was given these objects with a generosity never before seen. The two from Lambesc were of strong interest to me: these are for me new items that will one day be a topic for the history of the sciences. I consider these wonderful godsends as a kind of replacement for much lost time, that is, time I have not been employed in the silent research of the laboratory.

P.S. You know, My Lord Count, about my four stakes connected by a single rope. My lack of foresight! I didn't know the people of Provence. My device was like a feeble obstacle set before a brilliant cavalry charge: it was broken, cords and stakes, unbeknownst even to my giraffe handlers, their head in the clouds. So much for my good nature.

The same day, the head of the squadron, commander of the company of military policemen of the Bouches-du-Rhône, reported to the prefect that the giraffe had passed through Lambesc (figure 6.3).

May 27, 1827: While France was surrendering to the giraffe's charm, Reshid Akif Pasha's troops took the Acropolis. Athens fell into Turkish hands. At the time, the government of Charles X did not officially react, but the clan of Philhellenists was polishing its weapons.

May 28, 1827: The prefect of Marseille wrote to the Minister of the Interior the moment he learned that the convoy had left his department and therefore his jurisdiction. Under the rigorous tone of an organized public servant, we can perceive here a certain relief that the royal gift had been kept safe from any harm. Villeneuve responded quickly to Geoffroy's request; he had asked for intercession from a high place to justify a departure one might consider delayed:

> The various animals sent by the Egyptian Pasha for the royal zoo, and which spent six months during the cold season in Marseille, have been delivered to Mr. Geoffroy Saint-Hilaire, administrator commissioner of the Museum, who set them on their way the 20th of this month. Today I learned that this convoy has arrived at the outskirts of the Department, all in the state of preservation that I have had them maintained by continuous care. I informed the prefects of the Vaucluse, Drôme, and Isère, letting them know such details as to facilitate the passage and journey of the Giraffe, the most interesting item of this collection.
>
> While Mr. Geoffroy Saint-Hilaire arrived on the 4th of this month, it was not possible to leave any sooner. Several days first passed before study of the Giraffe's habits led him to be able to judge what type of preparations had to be made. We then had to wait for the workers to carry out the orders placed for construction. However, this necessary time was not lost to the sciences due to Mr. Geoffroy's actions.

Figure 6.3. The 1830 Almanac set the giraffe right in the midst of military police and the crowd.
© Aynaud Collection.

We note here the subtlety of the prefect who added, in the margins of his secretary's copy, so that no one would suspect him of indulgence with regard to the naturalist: "I owe him this account in return for his useful actions with regard to people who cultivate this type of knowledge within the department."

The letter continued:

> As a naturalist, he visited all public and individual collections and responded to those who wished to know the scientific names of a multitude of objects. He went to schools of anatomy and was brought to the Chair of the general hospital through unanimous solicitations by the city's doctors and secondary school students. There he gave lectures to update them on new developments in anatomical science, and his tireless kindness in this respect made an impression on this part of our population such that the results will only be fruitful; this has been so well-received that some expressions of satisfaction have reached me. The *Académie* in Marseille also wished to hear this knowledgeable professor and receive a talk on his point of view formed from various departmental information he was to collect.

Still more cunningly, the prefect concluded by transforming what Geoffroy feared would be judged as misplaced and inopportune—his visits to collections and schools—into an act of glory benefitting the minister himself: "In a word, this eagerness of Mr. Geoffroy Saint-Hilaire has been interpreted by several people as one result of his mission, and their gratitude has recognized the Ministry's intention for the development of all that might contribute to scientific progress."

On the same day, for good measure, the prefect sent the Museum a copy of his letter to the minister. Geoffroy would be pleased to have an ally of real diplomatic skill: "The nature of the details which I thought I had to record with regard to the circumstances of Mr. Geoffroy Saint-Hilaire's mission support my presumption that you would be interested in knowing them."

Not forgetting himself, he noted that he had given the naturalist a copy of his *Statistique*:

> As much in my name as in that of the departmental council. I hope that it will be of some use to you with regard to Science, but especially, that it will be a lasting token of all the satisfaction I have felt regarding connections that various circumstances have enabled me to forge with you.

Zarafa resumed her journey. In Avignon, notable local officials and friends of science escorted her through the district. In Orange, across from the amphitheater, she could not know that, in such a place, her ancestors were sacrificed in Rome for the people's amusement! Each afternoon, Geoffroy preceded the

CHAPTER 6

convoy to find and negotiate lodging for that evening. In La Palud, the female mouflon gave birth. A good sign, perhaps?

Then they reached Montélimar, Loriol, and Valence. In this last city, where Bonaparte was a cadet at the age of sixteen, a letter was waiting for Geoffroy.

June 2, 1827: Geoffroy Saint-Hilaire responded to the prefect, recalling an event that must have put him into a cold sweat:

> Your letter, a monument to generosity, that I will keep always, has reached me in Valence.
>
> Yesterday I left my convoy to spend the night in the city of Tain and came here [to Valence] by coach to prepare ahead for its arrival. I will leave tonight and go to join it on the road. Everything has gone as expected. The giraffe had picked up a nail in the membrane between her hooves. She did not limp. However, I noticed that she seemed fatigued by the length of the walk and that worried me. I slowed her walk and she will arrive in Valence in six days instead of four. She will spend the night two leagues from Lyon in Saint-Symphorien. I was thinking that I won't be able to escape from the solicitations of those curious onlookers once she enters Lyon. I will have her arrive without being fatigued by an entire day's walk.
>
> I have just organized an event for the curious of the great city of Lyon. The giraffe will be seen on the *Place Bellecourt*, among the trees of this magnificent square. It's better to foresee and organize in advance rather than be taken by surprise and lose one's head once the crowd arrives.

For the first time, Geoffroy established a plan that would allow for completion of a more peaceful trip:

> And so I came to Lyon to take stock of the place and means of the city, in the event that I may proceed with the idea of transporting her by water from Lyon to Châlons. And I found that this was possible and organized everything so that she could be embarked on the Saône River the 9th and would leave as of 4 a.m. She will arrive on the fourth day.
>
> You would not believe, my Lord, how her manners have improved. She is pliable, so much so that she is happy to be in perfect obedience. In Loriol, in the evening and on the following morning, she drank her bowl of milk very bravely before the entire company and she does it no other way today. But I find she is very tired, and it is better to go back to taking hygienic precautions rather than waiting to administer remedies to her. If I hadn't judged her transport to be possible by water, my plan would be to isolate her and keep her for a week in a remote country corner.
>
> I think that she will once again be fatigued upon her approach to Paris and I think that in this state one should not throw her into a fever of curious onlookers:

my difficulties in this regard are growing in proportion to the size of the population. I have written to the minister, that if the king so authorized it, we could, in passing by Montereau to Fontainebleau, have her rest for eight to ten days in our kings' palace. In addition, the royal family could organize a hunt there and be the first to see the animal.

June 5, 1827: The professors thanked the prefect for his letter of May 28th, for his care of the giraffe, his concern regarding Geoffroy Saint-Hilaire, and his gift for the Museum.

The same day, Geoffroy wrote to the minister. The tone of his letter was contrived, convoluted, and obsequious. Stress and fatigue had no doubt cost him his expression. It was surprising on the part of a man who had confronted the worst dangers in Egypt, then in Spain, and Portugal, but he was after all thirty years younger then:

My Lord,
A deep and well-deserved respect for the minister of a great kingdom makes me afraid to speak to your excellency about an affair of a disproportionate detail, considering all the great affairs that you direct; however, in asking your forgiveness for the necessity of sending you this report, I pray that your excellency, my Lord, will have the kindness not to be inconvenienced in the slightest.

The Giraffe continues to enjoy perfect health, and up till now, on my knowledge having arrived in Tain, in the Department of the Drôme, is perfectly well: she has sustained the fatigue of her journey courageously; from Marseille to Tain, sixty-six and a half leagues have been covered, by our measurements. The Giraffe has successively slept in Aix, Lambesc, Orgon, Avignon, Orange, La Palud, Montélimar, Loriol, Valence, and Tain until today; she will stay in Saint-Lambert tonight, tomorrow in Auberive, the following day in Vienne, and the day after that in Saint-Symphorien, to arrive without fatigue in Lyon on May 6th. I have just slowed her walk, I have noticed that while bravely withstanding all the fatigue imposed upon her, she has felt its effects nonetheless. The cows are starting to tire; one of them, born in Egypt, is limping a little. The Giraffe herself had picked up a nail in the membranes that link the two hooves, it has been taken out in time and before the animal had begun to limp; I was very worried about this accident that fortunately has amounted to nothing.

However, my Lord, I cannot hide the fact that the trip has in its duration brought with it an overall fatigue, a discomfort in all the animal's movements; having followed her until her night's lodging in Tain, I have set in my place and at the head of the convoy a doctor from the neighboring area, whom I knew as one of my former students and I went that night with the coach to Lyon to pursue two goals in the interest of the animal's health, confided to me under the patronage of

CHAPTER 6

your excellency: I will return by carriage at the front of the convoy for her arrival in Vienne, and then at the time agreed upon for Lyon.

Experience on the road has taught me that public curiosity, and consequently the fatigue of displaying the Giraffe, has grown in direct proportion to the size of the population. The animal is such a sensation that people wish to see her several times and that for second visits come all and sundry of the day before, along with those who were uninterested in the first news of her passage. I wanted to establish in advance with the mayor of Lyon some restrictive measures since this goal must be attained: the fullest contact with the population, under the condition and consideration that the animal be not at all affected by it. It is this middle ground that is difficult to comprehend and where I allow myself to interpret the gracious and generous sentiments of your excellency in every stop, which is for me a most serious affair. I think that in Lyon the means are attainable: My Lord the Count of ... wishes it to be thus; I believe the same is true for the mayor whom I have not yet had the honor of meeting in his office.

My other worry and stronger preoccupation was a necessary follow up to the remark I made as to the Giraffe's lassitude; eight days of rest appeared nothing short of essential to me, but a great impatience to see this animal arrive in Paris was also a consideration. I believed that a trip from Lyon to Châlons by water would fix everything, would give the animal eight days of rest without taking anything away from the number of potential days for its promenades, but it was crucial that such a desirable plan also be feasible and I wished only to speak for myself in evaluating the opportunity as to the places and supplies available; in addition, I have seen that the quays, the jetties, and the boats, with a bit of care taken for their appropriation, could be rather suitable. This avenue has become possible as the Giraffe is becoming more and more pliable and truly, through its docility and domestication, has adopted manners that compare with those of the Horse and the Camel; so, without fear of compromising her, I foresee the means of embarking and disembarking her and having her make the crossing on water from Lyon to Châlons; the crossing covers but a third more in distance than that of the route via land.

I think that the animal's health approaching Paris should be fortified by a stay of eight-to-ten days at some distance; there will be no rest possible upon her arrival if she is delivered straight away to the ardent curiosity of the capital. If the king would permit the animal to be received into the stables of his palace in Fontainebleau for a time, which would allow for the option of seeing the animal earlier by his majesty and the members of his august family, I would turn the convoy away from its scheduled route, I would lead it in the direction of Montereau toward Fontainebleau; it could be that the curious would come from afar before its arrival in Paris, which would perhaps constitute a bothersome crowd; a sudden change organized in the moment would prevent what is currently occurring, it is up to *monseigneur*, your excellency, to prescribe this; if, on the contrary, I were to receive no order, I would follow the road to Burgundy, as normally planned by the administration.

After disembarking in Châlons, I will submit the itinerary for the remainder of the journey to your excellency, and if authorized, I would do so with eagerness and punctuality.

I have the honor to be, My Lord, your excellency, a very humble and respectful servant.

Figure 6.4. Giraffe with two keepers.
© Aynaud Collection.

CHAPTER 6

June 6, 1827: Zarafa reached Lyon. It was a crucial point, justifying her four-day stay there. *La Gazette Universelle*, a local paper, quickly reported the information:

> The giraffe arrived today in our city where its lodging had been predetermined by the Provençal administration. She will stay here until the end of the week and will be walked several times along the strip of lawn that separates *Place Louis-le-Grand* from the *Promenade des Tilleuls* and along the pathway containing the flower market. The outings are tentatively fixed as follows: Wednesday at 5 p.m., Thursday and Friday at 10 a.m. and 5 p.m., and Saturday at 10 a.m. As her departure for Paris will be by water[!] that very day, precautionary measures will be taken to maintain order in the crowd of curious onlookers we expect may jump onto the giraffe's path; her itinerary is set and overseen by one of the members of the *Académie des Sciences*, Monsieur Geoffroy Saint-Hilaire. This scholar is authorized to ask authorities within places traversed for all facilities and assistance necessary for the safety and preservation of this beautiful animal entrusted to his care.[6]

After this, *La Gazette* published its daily report:

> The giraffe has had the greatest success in our city. For the past three days, a large crowd has assembled to wait for her in front of her lodging in Provence, and the moment she appears with her long legs and long neck, the cries made by all are deafening. In these cries there is a joyous note not always heard in popular assemblies. The giraffe has shown herself several times in public, preceded by a cavalry detachment, surrounded by a number of guardians who hold back the crowd. Four of the blackest men one can imagine walked by her side. They formed a court and contributed to her picturesque appearance. Finally, a member of the Institute accompanied the beautiful African lady.
>
> The African beauty does not appear at all surprised by the crowd rushing after her, she looks serenely at the curious onlookers from the height of her fourteen feet. She licks those who approach her with her enormous tongue.
>
> People who didn't know of the giraffe's arrival or perhaps who didn't know what a giraffe was, were at first surprised to see so many people assembled, along with armed troops. We heard a good woman inquire what all of the fuss was about. They responded to her: "The beast will appear." "The beast?" "Yes, the beast." And the good woman continued on her way, convinced that she was dealing with one of those people always ready to make insolent remarks concerning people of importance.[7]

Unfortunately, it was difficult to mobilize the entire police force, for "a great number of horses were grazing outside the city, and the rest were needed for the postal service." They therefore made arrangements using other means.[8]

In such a situation, an incident seemed inevitable:

> She has, however, fallen prey to a few moments of folly. When curious onlookers ran very quickly toward her, she became afraid and began to gallop around the statue of Louis XIV as she might have done in the deserts of Africa. The Blacks wanted to restrain her; three of them were thrown to the ground. Only one did not let go and finally managed to calm the poor animal. The guards in vain told the people not to run after the animal as they might frighten her, but they only ran faster. In any case she is certainly highly susceptible to emotions and the sight of the crowd running toward her put her into a nervous state. The horse of one of the hunters escorting the giraffe took fright hearing the cries of the crowd and brusquely stepped back despite the efforts of the horseman, knocking over a woman in her eighties; she was transported to the general hospital along with a young man who had been rather strongly pushed about in the melee.

The information would be relayed to the capital, in a slightly altered version:

> In Lyon, the enthusiasm was no lesser; it led to agitation. The Giraffe, walking toward the prefecture buildings, escorted by an immense audience of spectators, took fright at a barking dog; she disengaged herself from the long cords held by her keepers and ran in a circle. The crowd, ignorant of the quadruped's inoffensive behavior, but terrified by the threatening appearance of its colossal size, looked for an exit, clashed with each other and fell down; fortunately, one of the keepers quickly calmed the Giraffe's fright, and put an end to the accidents.[9]

The first third of its journey over, an initial report had to be made:

> There has been no change to her health, neither during her quarantine, nor during her stay in Marseille, which lasted about six months. She made it to Lyon without tiring herself, easily achieving two leagues [about five miles] in an hour and sustaining this walk as well as the best horses. She has been led by short day trips; this is due less to avoiding her fatigue than for providing her appearance to numerous groups spread across the countryside she has crossed. Everywhere people have rushed from afar to see her.[10]

Geoffroy Saint-Hilaire lost no time. The director of the veterinary school in Lyon offered him a real "curiosity": "a polydactyl monstrosity in a kind of horse." But he also gave of himself: during a brief moment of panic, the military policemen, leaving in a gallop, knocked over several people, including the naturalist, who was a victim of wounds "more numerous than serious."

CHAPTER 6

But it was the minister's silence that caused the greatest vexation. Geoffroy received no response, not even a negative one, regarding his proposal to travel via the Saône River. Disappointed, he wrote nothing more to His Excellency, taking up the long walk in Burgundy once more, along the former Roman roads, 120 kilometers [75 miles] of very straight passage between river and vineyards. Aside from his rheumatism, his gout and the wounds received in Lyon, Geoffroy suffered from an attack of uremia, which obliged him to travel sitting down more and more frequently.

The king did not appear unhappy with this *tour de France* that distracted his subjects at a time when his popularity was not at its highest, but he remained worried "that this jubilation might degenerate into revolt."[11] Was it fear of the people that kept the king from setting out to meet his gift, or rather respect for etiquette decreed by the intransigent *Madame Royale*? This Duchess of Angoulême, born Marie-Thérèse Charlotte de Bourbon, eldest daughter of Louis XIV and of Marie-Antoinette, was imprisoned for three full years in the Temple, exiled to Austria, then to England, married to her cousin the Duke of Angoulême, nephew and future successor of Charles X. At fifty years of age, this acrimonious, obese, childless ultraroyalist was reported to have said, "It is the giraffe that is to be led to the king, and not for the sovereign to run after it like some commoner before the gift given to him."

June 8, 1827: Before leaving Lyon, Geoffroy made an account to the prefect. With fatigue overcoming him, his prose no longer contained the lyricism of his journey's beginnings:

> I intend to have the Giraffe's convoy depart around 8 a.m. tomorrow in order to make it to Anse to spend the night; the second day, the convoy will sleep at the crossroads of the road to Belleville, spending five hours (midday) in Villefranche. The third day, we should arrive in Mâcon. Would you be so kind, My Lord Count, as to give orders for an escort of two to three military policemen and for more escorts within Lyon, until we have left the city?
>
> As prefect, would you also have the goodness to advise the Prefect of Saône-et-Loire regarding measures to be taken on the road, so that the military police of the Rhône department may be replaced by those from neighboring department brigades?

June 12, 1827: The Minister of the Interior thanked the prefect for his letter of May 28th, informing him of the giraffe's progress: "I desire only that the Friends of the Sciences harvest all the fruit for which they have been waiting." Geoffroy Saint-Hilaire would certainly have wished for less civility and more attention to his requests.

Even when outside the city, Zarafa continued to inspire journalists from *La Gazette de Lyon*, who tried to supply any new information regarding her trip: "The giraffe is continuing her journey. A traveler told us she had seen her last evening [June 15th] as she entered Châlons. She was in perfect health."

June 17th and 18th, the procession crossed the Côte d'Or.

June 20, 1827: The prefect sent his last expense report total of 1,861.50 francs. The sum represented more than three times the regular amount, as Geoffroy had wanted to embark grains in sufficient quantity to last until Lyon, and the employees had received their wages early in order to take care of certain expenses required for the trip. Among these additional costs, we note the purchase of a waxed canvas covering, harness and reins for the giraffe, and of a carriage. In addition, cart drivers in Marseille had to be reimbursed for damage caused when horse teams had taken fright at the sight of Zarafa.

Moreover, the prefect proposed to issue a tip of one hundred francs to his hotel's concierge, if the professors approved:

> It is for the man overseeing a daily watch of the animals' lodging, and for the foreign handlers. This watchman is in fact hired by the courtyards and stables providing housing; but he had to provide the most diligent care during the entire time the Arab keepers were lodged in these quarters and while a continual crowd of curious onlookers filled the avenues; I myself sought to reconcile the security and preservation of the giraffe with the rightful eagerness of the population and numerous travelers who asked to see her.

June 23, 1827: The elderly scholar's family was very worried about him. In Auxerre, Isidore Geoffroy Saint-Hilaire, then twenty-one years old, found his father along Zarafa's route. His father seemed relatively robust, although overwhelmed by his responsibilities. Isidore shared his anxiety with his mother:

> Between Joigny and Auxerre, my worry greatly increased: I set myself in the coupé alone—I, who never cry, and shed a flood of tears. For having found no authority in Joigny who had been advised as to the giraffe's arrival, I held the idea that Papa could not withstand the trip and had not made it to Auxerre. Shortly before I reached this city, I learned that the giraffe was there and soon after my arrival, I ran to the hotel where she was settled. Papa was not lodging there, but I learned right away where he was, and I was taken there. I found him in a room. He was pale but seemed fairly well....
>
> He had scarcely arrived, and we had to leave to dine in the city at the home of a man I did not know. They placed me at the other end of the table, far from Papa, and I suffered horribly the whole time: I felt ready to cry with regret as I was thus separated from him, at the very moment I had just seen him again, and by the

CHAPTER 6

thought that he would depart from me tomorrow morning, still suffering, and leave me in a state of worry, without anyone I could talk to about him and about you.[12]

June 26, 1827: Frightened by the fall of Athens, France allied itself with England and Russia in order to send an ultimatum to the sultan of Constantinople. Giraffe or no giraffe, the two European powers could not let the Ottoman Empire crush Greece.

The city chronicler in Joigny, Mr. Péri-Courcelle, recounted Zarafa's passage with a certain taste for comedy: "There she is, a boy cried, all of a sudden. And so she was. Marie Simpleton saw her at the end of the path, her head passing over the hedge."

But it was only a scarecrow at the end of a fishing pole, waved by a prankster. Finally, she did arrive, "in a good portly state, her body round and plump, her coat superb." A certain Mrs. Jeanniot had set a bale of hay on her second floor windowsill, so that the poor animal would have something nourishing as soon as she arrived. In fact, the hay offering suited the giraffe who, by pulling bits out with her mouth, pulled down the bale, which fell onto a townsman's top hat! Elsewhere, atop the mill in Montastruc (Lot-et-Garonne), as on the roof of the Chenoise castle farm (Seine-et-Marne), someone had installed a weathervane representing a giraffe led by its handler.

June 27, 1827: Approaching Paris, the crowd becoming ever more difficult to contain, Geoffroy Saint-Hilaire tried to establish a plan of entry. He wrote a confidential letter to the Museum, most certainly addressed to his colleague Georges Cuvier:

> My dear colleague,
>
> As a result of your fruitful imagining, or at least of your request to Mr. Drovetti, the giraffe is arriving to spend Friday and Saturday in Villeneuve-Saint-Georges. She is a superb animal that you will study with pleasure. We are not revealing the time of the giraffe's arrival to anyone. It's up to you, my dear colleague, to find the means to see her before [her] arrival. She will enter the king's garden on Saturday night.
>
> She must take an unannounced route to outwit everyone else: from Villeneuve, she will go to Choisy, where she will pass over the Seine. She will proceed from Choisy to Vitry on Saturday morning, where she will stay until we determine when she will make the last part of the trip. Ever since Châlons, I have suffered from a serious illness brought on by fatigue. My son has come to join me in Auxerre and will take over management of the convoy.
>
> My dear colleague, know that you have my greatest affection.

7

TRADE AND GIRAFFOMANIA

Marketing Materials

In 1797 in Épinal, Jean-Charles Pellerin launched a great adventure. He started a factory using printing techniques of his time, wood etching to express features, and stenciling to bring the image to life in color. In his small workshop on Léopold-Bourg Street, he gave birth to Épinal's future prestigious image printing company. In 1820 he adapted to using modern techniques, principally lithography. The small company experienced its moment of glory thanks to popular distribution, assured by merchants who traveled from town to town, hawking the factory's products. The giraffe was a golden item! (figure 7.1).

In Maisse (Essonne), a ring was sealed in the inn's walls, thereafter baptized *Hôtel de la Girafe*. In Tonnerre (Yonne), an inn took as its sign *"A la giraffe"* until the beginning of the twentieth century. And this was only the beginning of a giraffomania that would pervade all branches of commerce (figures 7.2 and 7.3).

June 28, 1827: Zarafa passed through Melun and then Fontainebleau. French author Stendhal (né Beyle) was getting ready to take a steamboat up the Seine River to encounter this *grande dame* in Villeneuve-Saint-Georges, but the procession was delayed twenty-four hours. He wrote to the minister of Tuscany in Paris, Deniello Berlinghieri (in Italian):

> Excellency, great figures are subject to a change of mind, as you know. I have just received word at 5 p.m., that the very tall giraffe will not arrive in Villeneuve-Saint-Georges until Saturday. Thus, on Saturday, at 7, we will leave, taking the steamboat by the riverbank. Mr. Geoffroy, this learned person, this great professor, has made the error of mistaking Friday for Saturday. I remain, with all due respect, your very humble servant.
>
> H. Beyle

Figure 7.1. The fashionable giraffe.
© Collections of Musée du Domaine départemental de Sceaux. Photo Gilles Vannet.

Figure 7.2. Engraving board.
© Collections of Musée du Domaine départemental de Sceaux. Photo Gilles Vannet.

Figure 7.3. Tapestry.
© Collections of Musée du Domaine départemental de Sceaux. Photo Gilles Vannet.

Another reason for this celebrated author to take a trip up the river seems to have been the presence of the charming Sophie Duvaucel, Cuvier's stepdaughter.

"Cuvier and several scholars were going to wait for her at the gate of Villeneuve-Saint-Georges, accompanied by twenty-five military policemen."[1]

June 30, 1827: Finally! The giraffe "has arrived at 5:30 p.m. at the Fontainebleau gate! She was led to the botanical garden by way of the hospital boulevard."[2] Zarafa had covered the 880 kilometers separating Marseille from Paris in forty-two days, seven of those as days of rest, versus the fifty-two days in total foreseen by Geoffroy:

> But it was mainly the giraffe who had so marvelously benefited from it. She had gained weight and was strengthened by the exercise. Her muscles were firmer, her coat softer and more lustrous upon her arrival in Paris than at her departure from Marseille. Now she is 3.8 meters tall. In addition, her manners are more self-assured. She no longer refuses to drink milk in front of strangers, and her indulgence in the little mouflon's games, allowing him to climb on her back while she lies at rest, attests to her good nature as well as her intelligence.

As for the aging naturalist, he had to run to a surgeon to take care of an emergency related to urinary retention.

July 2, 1827: Assuming his obligations as a man of science, Geoffroy delivered an initial speech to the *Académie*. *Le Moniteur Universel* published an account of it four days later:

> Mr. Geoffroy Saint-Hilaire shared some curious information regarding the giraffe from Sennar, which arrived Saturday evening at the *Ménagerie*. He said that this giraffe was of a different species than the one from Cape Town, that we may see stuffed in the office of the *Histoire naturelle*, and he deduced the physiological proof upon which his opinion is based. He then gave the floor to Mr. Mongez, to read a history of the giraffe. That scholar began by carefully enumerating various documents he had gathered on this animal, from noted historians as well as laypersons....
>
> But what should really pique Parisian curiosity is that they have not been brought to Europe since 1486, and that the one we can now see in the *Jardin des Plantes* is the first living specimen to arrive in France [figure 7.4]. She was cared for on her journey by a slave from the dey of Algiers, who was responsible for another one sent five years ago to Constantinople, where she died shortly after her arrival.

CHAPTER 7

Figure 7.4. In the rush to see the giraffe, watch out for pickpockets. *Les Quartiers de Paris.*
© Musée Carnavalet—Histoire de Paris.

The next day, it was the giraffe who *wrote* to the newspaper, *Le Figaro*:

> Since my arrival in France, I have learned that you have spoken about me in the most flattering way; as I am sensitive to this gallant as well as disinterested behavior (for in praising me you were surely not counting on me for a subscription as I don't read French), I promised myself to express my full gratitude, and I am taking advantage of a moment of rest to keep my word. Perhaps you will find my action a bit unceremonious: I should have sent you my card or gone in person to assure you of my gratitude, but unfortunately, I have neither servant nor card to send you, and it is impossible for me to move about, considering that my new hosts have developed such an attachment to me, that they do not wish me to leave their establishment, which has led me to decide to write to you.

TRADE AND GIRAFFOMANIA

I will admit to you, Mr. Figaro, that while I have been pleased with your praises and your punches, which have prompted my hearty laughter, I have not been able to endure the singular comparisons to which you have submitted me without making my acquaintance. At times you have stated that I have had to live with one you likened to fire tongs, or that my appearance has comprised a set of grotesque elements whose shapes you have taken care to enumerate.

If I were as bizarre as you have judged me to be, I might wish to lodge a dispute with you, like the Chief Editor of the *Constitutionnel,* or ask for compensation for your insults after having embraced you for your flattery. But fortunately for you, I do not wish to do as others do. I will not propose a duel because, decidedly, I cannot pay for the lunch you would expect, as I believe that we will not eat from the same hayrack. Therefore, Mr. Figaro, we will remain good friends, and to seal our friendship, I ask you for a favor that will be easy for you to render to me: it is a matter of thanking in my name all the scholars who have taken the pleasure of devoting their care to me, during my travels, and since my arrival.

I hope very much to receive a visit from you and I dare to hope that you will want to honor me with one.

Your very humble servant, *La Girafe*

July 4, 1827: The professors, finally reassured, thanked the prefect:

We are eager to announce to you that the Giraffe has happily finished her journey, and that she has arrived in the King's garden on the 30th of last month, without having incurred the least accident, and remains in the best possible state.

We think that we owe in great part these happy results due to the care you took to accustom the animal to the noise and sight of the crowd from the first moments of her arrival, and we have the honor of expressing once again all our gratitude.

Her visiting hours were published as far as Lyon: "As of Monday, this animal, any unforeseen incidents notwithstanding, will be walked every day, weather permitting, from 10 a.m. until noon within the botanical school, a vast enclosure surrounded by metal railings around which the public will be situated so as to see her without obstacle, and free from danger."[3]

On the first day, more than ten thousand people visited her, and from July to September, she would attract six hundred thousand visitors. The naturalist Bory de Saint-Vincent, locked up for his debts in the prison of Sainte-Pélagie, had asked to see her but was refused this exemption. His friends from the Museum took advantage of one of Zarafa's walks to pull her toward the small Labyrinth hill, near the Jussieu cedars. And Bory, from the roof of his prison, could then at least see her through his telescopic spyglasses.

CHAPTER 7

Figure 7.5. Giraffe with Diogenes. "Diogène avec sa lanterne cherchant un homme en plein midi."
© Aynaud Collection.

July 6, 1827: The giraffe had a few critics, as one always does. One journalist from *La Gazette de France* commented:

> If an animal's beauty consists in the proportions of the parts that constitute it ..., in the suppleness and harmony of its movements, in its force of character, speed, dexterity, or in its type of strength, the giraffe is not one of the most beautiful animals. She is one of the ugliest, the undisputed ugliest of the large quadrupeds.... Her long neck, which seems rather flexible, has nothing in common with the camel aside from its length, which appears still more disproportionate than that of the camel. Her head set atop this neck seems all bone and somewhat resembles an ostrich.... If by her mass, her bearing, her behavior, the giraffe is comparable to something, it is less to an animal that we know, than to a kind of theatrical model like the one used for camels in *La Caravane*, whose dummy is carried by a single man, the neck of the animal portrayed by nothing but a stick covered by a canvas sleeve.

July 8, 1827: La Gazette de Lyon announced that "a new lithography of the giraffe, representing her in a very detailed and lifelike manner would appear at the beginning of the week. The drawing by Mr. Flandrin would include a small plaque at its base."

From that time on, the image of the giraffe would be found on every kind of item imaginable, to the great benefit of merchants: dishes (dinner plates, serving plates, bowls [figure 7.6], powder shakers, vases, cups, salt cellars, carafes, vials, terrines, basins, washbowls, pitchers and jugs), earthenware (from Nevers, Gien, Montereau, Clairefontaine, Les Islettes), porcelain (from Delft, Limoges, Paris), glass panes, statuettes, toothpick holders, cigar cutters, candlestick holders, matchboxes, snuffboxes, candy boxes, flat irons, bedwarmers, medals, firebacks, almanacs, bronze items, pendulums, signs, weather vanes, tapestries, combs, inkwells, paperweights, lamp stands, toys, gingerbread, cake pans, giraffe-style sauce, toys with articulated necks, bedcovers, wedding armoires, soaps, wallpaper. Paris almost adopted giraffe lamp posts with Lebon natural gas to replace the old oil-based streetlamps.

Zarafa was in fashion. She dictated and dominated fashion: hats, belts, embroidered handbags, purses made of pearls, canes, fans, handkerchiefs, dance

Figure 7.6. Salad bowl. Saladier à bord dentelé.
© Musée Barrois Collection. Photo Michel Petit.

CHAPTER 7

cards, parasols.... The *Journal des Dames et des Modes* reported that its readers were completely captivated by this unavoidable motif: "Every lady decorates with a giraffe adorned with little spots, one has a woolen washcloth or a screen, another a cushioned stool."

A giraffe necklace was "a narrow ribbon on which hung a rose pulp heart, or better yet, a seraglio medal in the shape of the amulet seen around the giraffe's neck at the *Jardin des Plantes*."[4]

A hair bun was put up *giraffe-style*, vertically with ribbons, knots, feathers, and flowers. For men, waistcoats, hats, ties, and high collars were known as *giraffic*. Clothing was sold under the names *Giraffe Belly*, *Amorous Giraffe*, *Exiled Giraffe*, *Pale Savannah*, or *Sudanese Marble*.

Her grace and elegance, her gentleness and dignity, were praised, her eyes admired: the highest compliment one could pay a lady was that she had the eyes of a giraffe! Invented in 1819, the claviharp was renamed the *piano-girafe*. A number of songs were written, such as a waltz by Singer for piano, and a light guitar piece by F. Carulli, "Les Adieux de la girafe."

And this "Cantata for Choirs," set to the tune of "Cadet-Roussel":

Déjà sur vos brillants attraits
La mode a basé des succès;
Sur les foulards, sur les gilets,
Partout on reproduit vos traits;
Et coiffure et robe nouvelle
Tout se taille à votre modèle.

[Already upon your features bright
Fashion set its success on this height;
On waistcoats, on scarves unfurling,
Everywhere your features spring;
And in hairstyle and new dress,
All is tailored giraffesque.]

Solemnly titled *La Girafe*, this printed booklet must have been sold around the King's Garden:

C'est de l'acacia qu'elle aime à se nourrir
Mais la liqueur du lait fait son plus grand plaisir
L'Africain, envieux de sa robe tigrée,

Lui lance dans le flanc une flèche acérée;
Le lion moins cruel respecte sa douceur:
En tout temps, elle fut présage de bonheur.
Et lorsque Constantin sous son sceptre en eut une,
Rome y vit une grande fortune.
Les Hottentots encore, heureux de son trépas,
Forment de ses désirs leur plus exquis repas
Enfin dans tout Paris, on aime sa présence
Et son séjour prédit la paix et l'abondance.

[From the acacia she likes to eat
But a milky drink is her greatest treat
The African, desiring her tiger-like hide,
Shoots a biting arrow into her side;
The lion, less cruel, respects her sweetness:
In all times, she foretells happiness.
When Constantine had one under his scepter,
Rome saw great fortune come to her.
The Hottentots as well, happy to kill the beast,
Made of her pleasures their most exquisite feast
And finally, all across Paris, one loves her presence,
She in residence, a sign of peace and abundance.]

In Paris and Marseille, Adolphe Jauffret published *Trois Fables sur la giraffe* [Three fables about the giraffe]:

La Giraffe
On vous a vu, Monsieur le Fabuliste
Suivre de la loin, la Giraffe à la piste.
Cet animal, Léopard et Chameau,
Pour vous, sans doute, est un acteur nouveau.
Fier d'un bonheur que n'eut pas La Fontaine,
Vous voulez mettre une Giraffe en scène...
Oui, j'y rêvais; je ne m'en cache pas.
Quand sur nos bords, elle marche à grands pas;
J'aime à lorgner cette beauté d'Afrique
Type vivant du genre romantique,
Son seul aspect est fait pour étonner.

CHAPTER 7

Mais quel emploi, quel rôle lui donner?
Un premier rôle, un rôle de Princesse
Courant le monde, et, dans tous les pays,
Fixant les yeux des peuples ébahis.
[. . .]
Paris veut voir la Giraffe à son tour;
Peut-être même ira-t-elle à la Cour . . .
Eh bien! Paris, la Cour, lui feront fête.
Qu'elle s'y rende et suive son destin!
La Capitale est un séjour divin.
Là, plus qu'ailleurs, que l'on soit homme ou bête,
Qui vient de loin, et porte haut la tête,
Est assuré de faire son chemin.

[The Giraffe
We have seen you, Mr. writer of fables,
Follow her from afar, as she left her stables.
This animal, leopard and camel,
For you, no doubt, is novel.
Proud from a happiness not of La Fontaine,
You'd like to put a Giraffe on stage then . . .
Yes, I was dreaming, this I don't hide;
When along our banks, she walks with great strides;
I love to spy upon this African beauty,
Living specimen of romantic poetry,
Her appearance alone gives one a start,
But what employ, what role do we impart?
A premier role, one of Princess,
Roving the world, and in every country around
Fixing her gaze on those she astounds.
[. . .]
Paris wants to see her there also;
Perhaps to the Royal Court she will go . . .
Well then! Paris, the Court will give her a party.
May she render herself to her destiny there!
The Capital's residence is divinely fair.
Whether one be man or beast there, aye,
Those from afar do raise their head high,
and most certainly do make their way there.]

La Giraffe et l'Antilope
La Giraffe, en secret, disait, hier matin,
À l'Antilope, sa compagne:
Ma bonne, nous allons nous remettre en campagne;
On nous mène à Paris, et nous partons demain.
Je crains, s'il faut être sincère,
Que mon cou de Chameau, ma robe de Panthère,
N'y fassent pas fortune—on en raffolera.
Je marche vite, et je m'en pique.
Mais quand je cours, j'ai l'air de boiter. La critique
Saisira ce défaut, et Paris en rira.
Au contraire, quelqu'un dira:
C'est une allure romantique!
Et tout Paris applaudira.
Quand on verra ce cou, rival des cous des Grues...
Un concert de bravo montera jusqu'aux nues.
Oui; mais... —Que pouvez-vous appréhender encor?
Je bois, vous le savez, du lait.... C'est mon délice;
Et je mène avec moi la Vache ma nourrice.
On dira que je suis... —Digne de l'âge d'or.
À vous entendre donc, nous serons bien reçues...
Des flots d'admirateurs vous suivront dans les rues!...
[...]
Nous partons, Paris nous appelle.
La foule va m'y suivre, écarquillant les yeux.
Mais dans ce pays merveilleux,
Ma vogue se soutiendra-t-elle?
Que me demandez-vous! Toute vogue a sa fin.
Au Grand-Caire, à Stamboul, à Paris, à Marseille,
Souvent le peuple ingrat et vain
N'attend pas jusqu'au lendemain
Pour plonger dans l'oubli l'idole de la veille,
J'ai donc raison d'être en souci.
Les grands ont, ici-bas, bien des revers à craindre.
J'en conviens; mais hélas! s'il n'en était ainsi
Les petits seraient trop à plaindre.

[The Giraffe and the Antelope
The Giraffe secretly said yesterday

CHAPTER 7

To the Antelope, her good company:
My good one, let's go to the country;
We're called to Paris, and we leave with the day.
In all sincerity, I do fear
That my Camel's neck, my coat of Panther,
Will not gain fortune there—but they will adore you.
With fast walking I am piqued;
And when I run, I seem to be limping. The critique
will note this fault, and all Paris laugh too.
To the contrary, I will hear someone saying:
It's a gait that is so *romantique*
And all Paris will be applauding.
When that neck is seen, rivaling the cranes' . . .
A concert of *bravos* will mount to the heavens.
Yes; but . . . —What can you still learn from my sight?
As you know, I drink milk. . . . It is my delight:
And my Milk Cow nanny I'll bring along with me.
They will say . . . —Worthy of the golden age.
To hear you speak, we will be well received . . .
Crowds of admirers will follow you in the streets! . . .
[. . .]
We will depart, Paris calls us along.
The crowd will follow me, their eyes open wide.
But in this marvelous countryside,
Will I remain in vogue for long?
What do you ask of me? All trend has its end.
From Great Cairo to Istanbul, to Marseille, and to Paris,
Often the people ungrateful and vain bend
They do not wait until morning comes again
To forget their idol of the day before they don't see,
So then I do have cause for worry.
The grand ones here have many setbacks to fear.
I understand; but alas! If it were not this way
The littler ones would be so pitied here.]

La Giraffe et sa Nourrice
La Giraffe, rentrant à la ménagerie,
Sa fidèle Nourrice en ces mots lui parla:
Savez-vous ce qui se publie?

Un savant vient de faire, en pleine Académie,
Votre panégyrique. —On m'a conté cela.
Je m'en tiendrais fort honorée;
Mais, dans ce même discours-là,
Je n'en suis que trop assurée,
À l'Autruche on m'a comparée.
Le bel éloge que voilà! . . .
[. . .]
L'orgueil est chatouilleux. Quand ce philtre vanté
Qu'on nomme la louange aux grands est présenté,
Il y manque toujours à leur gré quelque chose.
S'ils ne peuvent tout haut se plaindre de la dose,
Ils murmurent tout bas contre la qualité.

[The Giraffe and the Nanny
The Giraffe heard, while returning to the zoo,
These words spoken by her loyal Nanny;
Do you know what they write, dear *amie*?
A scholar has just made your eulogy
to the *Académie*. —They told all this to me.
I would be most honored to hear
But in that same speech where
I have been most assured, my dear,
That to the ostrich I have been compared
What fine eulogy there! . . .
[. . .]
Pride is sensitive. When such an extolled potion,
Called praise of the great, is sold as a notion,
There is always something missing, by their will imposed.
If they cannot complain as to the portion's dose,
They still grumble about the quality of the mention.]

The press also ventured a bit of critique (figure 7.7):

All these giraffe styles are not in the best taste. Yesterday, at 9 p.m., several workers coming out of dance halls in Ivry had a rude quarrel, and one of them said to his colleague: "Send him a giraffe-style slapper."

"But who even knows what a slapper *à la giraffe* is?"

"I'll show you." The man encouraged in this way dealt a vigorous blow with his fist to his adversary's head, leaving him to the wine merchant in the horse market to bring the sorry chap back to life.[5]

Figure 7.7. Giraffes in fashion. *Encore des Ridicules*
© Musée Carnavalet—Histoire de Paris.

8

POLITICAL CARICATURE
A Satirical Weapon

Caricature provided a third, formidable way to utilize Zarafa's presence in France, in addition to scientific illustrations and commercial exploitation. Satirists would fully embrace this medium in order to denounce Turkish-Egyptian repression of Greek efforts for independence, to fight against the reactionary regime of Charles X and his Minister Villèle, and to oppose their king and minister's clericalism along with its quest for censorship.

Upon his return to France thirteen years earlier, Charles X, still the Count of Artois, was thought to have said, "Nothing has changed in France, there is just one more Frenchman." Then his rabble-rousing witticism was countered by the opposition, which engraved a new version of it on a medal displaying a giraffe's silhouette: "Nothing has changed in France; there is only one more big, dumb animal." The identification of Charles X with the giraffe was made clear by the choice of slender, elevated height to depict the sovereign, along with his prim and proper look.

One satirical sketch was titled, *La Girafe ou le gouvernement des bêtes est un divertissement impromptu donné par MM. Les Animaux du Jardin du Roi comme un témoignage de leur reconnaissance envers le Pacha d'Égypte, à l'occasion de l'arrivée de la Girafe à la Ménagerie de Paris* [The giraffe, or the government of animals, an impromptu comedy given by the animals of the King's Garden as expression of their gratitude to the pasha of Egypt, on the occasion of the giraffe's arrival at the Paris Zoo]. This satirical piece (cf. figure 8.1), published without authorial attribution in 1827, includes an excerpt of a speech on the budget by Chateaubriand in its epigraph. Disseminated under the table to avoid scrutiny, it was notably composed by the head of the opposition, Chateaubriand himself:

CHAPTER 8

Figure 8.1. Metamorphoses of the day.
© Aynaud Collection.

Scène 1
LE BAUDET
(Ministre de l'Intérieur, ayant le Département des Beaux-Arts)
Apprenez qu'un pacha, mes amis, quelle gloire!
(À tel excès d'honneur voudra-t-on jamais croire?)
Un pacha nous envoie du fond de ses déserts
Un animal qui marche sans bride et sans fers.
LE SINGE
(fonctionnaire du ministre)
Je doute qu'avec nous jamais on le compare.
UN AUTRE SINGE
Quel présent dangereux . . .
LE BAUDET
La frayeur vous égare;
Des gendarmes nombreux le suivent pas à pas,
Et d'après leur consigne, ils ne le quittent pas.
Enfin, pour aujourd'hui, l'agile télégraphe

À notre impatience a promis la giraffe,
Et c'est pour dignement ici la recevoir
Qu'une fête brillante aura lieu dès ce soir
[...]
Scène V
Divertissement offert par les Dindons (symbolisant le «parti prêtre»,
hostile à la cause de l'indépendance grecque)
... Oui, que tous les dindons que l'on engraisse en France
Viennent en choeur ici célébrer le Turban.
CHOEUR GÉNÉRAL DES DINDONS
Dindons, dindons, en danse,
Célébrez le Turban
Le Coran
Le Caftan
Le Divan
Le Sultan

[Scene 1
THE ASS
(Minister of the Interior, responsible for the Department of Fine Arts)
Learn about a pasha, my friends, what glory!
(To what excess of honor will one go with the story?)
A pasha sends to us from the depths of his desert,
Walking without bridle or horseshoes, an animal present
THE MONKEY
(Minister's public servant)
I doubt that anyone will ever compare it to us
ANOTHER MONKEY
What a dangerous present ...
THE ASS
Fright is leading you astray;
Many military police follow its every movement,
Advised not to leave it for even a moment.
Finally, today the agile telegraph responded
To our impatience, the giraffe is promised
And with honor we will receive her tonight
This evening here, with a party, festive and bright
[...]

CHAPTER 8

Scene V
Comic scene offered by the Turkeys (symbolizing the "priest's side," hostile to Greek independence)
... Yes, may all the turkeys fattened in France
Come here in a chorus to celebrate the Turbans.
GENERAL CHORUS OF THE TURKEYS
Turkeys, turkeys, in the dance,
Celebrate the Turban,
The Koran
The Kaftan
The Divan
The Sultan]

July 4, 1827: Indifferent to the strong emotions she inspired, Zarafa continued her mission, as noted by the *Moniteur Universel*:

> The celebrated giraffe arrived a few days ago in Paris. A negro from Darfour named Atir, and a Moor from Sennar named Hassan, both sent by the Egyptian pasha, and coiffed with turbans, held the animal on a leash, and were followed by two other Africans [Joseph and ?]; at the gate, her traveling outfit was removed, composed of a waxed canvas bearing the French coat of arms; an escort of twenty-five military police had been sent to the gate of Villeneuve-Saint-Georges. During the entire route from Marseille to Paris, three military police have been successively provided by every station, in order to protect her from meddlers. A carriage containing various other animals sent by the pasha to the king of France, preceded the procession; then we noted Mr. Geoffroy Saint-Hilaire, who having neglected his own health in the interest of science, had accompanied her without interruption until a short distance from Paris, and had only given her over to his son's care when his illness had become serious enough that further efforts on his part would have endangered his life.
>
> The giraffe has been housed in the *Orangerie* with several other Egyptian animals sent with her.
>
> It was yesterday, Monday, that she made her first promenade. More than ten thousand people have successively gone to visit her. The public is allowed to see her daily from 10 a.m. until noon.

July 6, 1827: Current events caught up to Zarafa and her ambiguous status as gift. From the start of the Greek struggle for independence, European sympathies had been displayed, through Philhellenism, via the works of poets and other artists, or in the courage of volunteers such as Byron. Paralyzed at first by the indecision of Alexander I, and the calculations of Metternich, the European powers then took action. Urged to unite for various reasons, whether by enthusiasm, interest, or fear, France through Minister Villèle, Russia through Nicolas I, and England through Minister Canning were brought together in the Treaty of London to impose their mediation.

Around that time, a link between France and Algeria would become official. For three centuries, under the designation as Regency of Algiers, this city and the land of present-day Algeria were then under the theoretical sovereignty of the Turkish sultan of Istanbul. In fact, rebellious and resistant to Islamification, the country's interior was abandoned. In 1798 the Directory government purchased wheat from the Regency of Algiers for the needs of Bonaparte's expedition into Egypt. These supplies were financed by a loan to France made by Jewish families in Algiers. They asked for a guarantee from the dey governing the city.

But in 1827 Hussein, dey of Algiers, tapped the Consul of France Deval with the handle of his flyswatter; Deval was a wheeler-dealer who had insolently refused to provide guarantees for reimbursement of the debt. Charles X saw an opportunity there. Six French warships quickly repatriated the consul along with other French nationals.

Villèle, president of the French ministry, demanded reparations to the dey for offending the consul, but he did not receive anything resembling a request for pardon. Proceeding from faux-pas to misunderstanding, this ridiculous imbroglio would spark a long and brutal conquest and a colonial tragedy that continues to wound France and Algeria even today.

July 7, 1827: Messieurs Théaulon, Anne, et Gondelier created a vaudeville play, presented for the first time in Paris in the theater of Vaudeville, called *La Girafe, ou une journée au Jardin du Roi* [The giraffe, or a day in the King's Garden], set to popular tunes now lost to the past:

Mme Bertholin (Air de M. Gaspard):
De son voyage pour Paris
On a donné l'itinéraire
Et le maire en chaque pays
Visita la belle étrangère;
De gens partout on l'entoura.
Pour mieux dissiper ses larmes. . . .
À Lyon, même, on lui donna
Une escorte de vingt gendarmes.

CHAPTER 8

Bertholin (Air du Vaudeville de la visite à Bedlam):
Ici tout Paris se rend.
Quels caprices plus bizarres!
Les grandes bêtes pourtant
À Paris ne sont pas rares.

Robert (Air de la sentinelle):
Sur leurs remparts, le mousquet à la main
Ces nobles fils de la Grèce asservie,
Pour relever quelque jour son destin,
Aux Ottomans disputent leur patrie.
En vain par le fer des combats
Un tyran moissonne ces braves;
La terre de Léonidas
Enfantera bien des soldats
Avant d'enfanter des esclaves.

[Madame Bertholin (tune by Mr. Gaspard):
During travel up to Paris
With her itinerary clear
The mayor of each place
Visited the stranger dear;
People, they encircled her
Her tears to dissipate....
In Lyon, they even gave her
An escort of twenty police from the state.

Bertholin (tune of the Vaudeville from Bedlams' visit):
Here all Paris runs to her,
So capricious and so curious
Large animals, however,
In Paris are notorious.

Robert (tune from the sentinel):
On their ramparts, musket in hand
In captive Greece, those noble sons
One day raised their destiny grand,
Disputing their fatherland with Ottomans.
In vain through the iron of battles
A tyrant harvests these sons brave;

The land of Leonidas
will give birth to many soldiers,
before those who would be slaves.]

Le Figaro published a mixed review of this show in its July 9 edition:

> The giraffe has inspired this light piece whose authors certainly don't give it much weight. A student disguised as tour guide in the botanical garden flirts with the daughter of Madame Beastophile, a tall woman decked out in a tiger-like dress, coiffed with a horned bonnet, and happy to note this analogy with the giraffe; there are two young people disguised as Egyptians in order to evade pursuit of their creditors, and whose uncle pays their debts. The giraffe who is scarcely mentioned in the whole play appears at the end, not in flesh and blood, but in cardboard, moving neither legs nor arms.

One text, *Dame Girafe à Paris, ou Aventures et voyage de cette illustre étrangère, racontés par elle-même, en réponse au discours de S. E. l'Ours Martin; avec le détail des fêtes que lui ont données les pensionnaires du Jardin du Roi* [Dame Giraffe in Paris, or the adventures and travel of this illustrious foreigner, recounted by herself, in response to a speech by His Excellency the Bear Martin; with mention of the festivities that the pensioners at the King's Garden gave her], is an *à-propos historique* [timely historical piece] by Charles-François Bertu, *précédé d'une dissertation scientifique par Buffon à l'usage des visiteurs de la Ménagerie* [preceded by a scientific report by Buffon for usage by the zoo's visitors]. Exhibiting an explicitly Philhellenist perspective, this text recounts the story of a Greek girl who was picked up by the giraffe's boat, having escaped from the Turkish massacres:

> I don't know which unfortunate people threatens the sovereign who rules in this place, but I have witnessed vast preparations for war, and at the moment when the four Egyptians accompanying me pulled me toward the ship ... I heard a man with a wild gaze, his head crowned with a turban, say, "They sent us cannons, our war vessels are being built on their construction sites; let us send them this animal. Presents maintain friendships." And he added with pride and fury, "We are conquerors of these odious islanders who, daring to refuse a slave's tribute to their master, have raised the flag of independence against him. By Mohammed, they all perished! Chio, Missolonghi, have fallen under the bronze avenger of the Ottomans, this city so dear to them, this city of which they are so proud, and almost under our power; with the European saltpeter, we will blast the last columns of Minerva's temple."

CHAPTER 8

L'Épître à la Girafe [The epistle to the giraffe] of a certain Charles D.... is equally explicit:

Pardonne, ô jeune prisonnière!
Je pleure un peuple entier, je pleure des héros
Qui n'ont pas un maître, ils n'ont que des bourreaux.
Esclaves comme toi, mais cent fois plus à plaindre,
De leurs cruels tyrans, les Grecs devaient tout craindre....
Nous avons un pays, des enfants à défendre
Courage, les Chrétiens viendront nous secourir
À l'oreille des rois leur voix s'est fait entendre
Les rois les ont laissés mourir!

[Have pity, oh captive young lady!
I cry for an entire people, I cry for the heroes
Without master, only executioners to follow.
Slaves like you, but to pity one hundred times more,
The Greeks must fear the worst from their cruel conquerors....
We have a country, and children to defend
Courage, Christians will come to our aid
To the kings' ears their voices send
The kings have left them all for dead!]

July 9, 1827: In his castle in Saint-Cloud, Charles X impatiently awaited his gift, unhappy to be the last of the French to approach the fabulous beast, and would happily have gone out to meet her and be presented with *his* giraffe, if it were not for a veto given by the Duchess of Angoulême! For Étienne Geoffroy Saint-Hilaire, after a courageous month and a half of joy and torture, it was a painful apotheosis. As for Zarafa, it was the French consecration of a beautiful Egyptian orphan (figure 8.2).

La Dernière Notice [The latest report] gave details on this historic day:

> For the first time, she traveled without her three Egyptian cows, her companions and wet nurses; crowned with a garland of flowers, the amulets from Mecca around her neck, she walked as in triumph, led by her guardians, and under the protection of a military police detachment. Several scholars and distinguished figures preceded her and followed her on horseback and in carriages; a crowd of curious onlookers who had rushed in from neighboring villages and countryside crowded along her path.

Figure 8.2. Last report on the giraffe containing the story of her trip to Saint-Cloud. *Dernière Notice sur la Girafe.*
© Musée Carnavalet—Histoire de Paris.

Having arrived at the park gate, after a three-hour walk, she was led up to the Trocadero. The king and his royal family appeared soon after. The elevated Giraffe was rendered even more gigantic and striking in the eyes of the noble figures.

The king and heir apparent, the Duke of Bordeaux and the princesses, continually admired her contours that were altogether bizarre and elegant.... A ceremonial cloak of red and black was thrown over the Giraffe's body, then they removed this artificial decoration to let her display the velvety, tiger's coat she wore naturally....

His Majesty wanted to see the animal run. They urged her to do so, and she made three rounds of the square at a trot, throwing first, as already stated, her forelegs out, then bringing her hind legs to a foot from where she had started. After this exercise, *Mademoiselle* presented her with a bouquet that she graced lightly with her delicate mouth. The King himself was pleased to have her eat from his hand, some rose petals of which she appeared to be very fond. The caresses of this animal, her gentleness, and affability deeply touched the Princes and all members of the Court.

CHAPTER 8

Le Moniteur Universel sent a special dispatch:

> Yesterday, at 10 a.m., the giraffe, who had left at 6 a.m. from Paris, arrived in Saint-Cloud, where she was led by the Trocadero to the Orangerie. A great crowd of onlookers accompanied her.
>
> Representatives of the Institute, composed of *Messieurs* Cuvier, Geoffroy Saint-Hilaire, and all members of the administration of the King's Garden presented this animal to the king, explaining her character and habits to him. This group had the honor of being presented to the king before mass, by His Excellency, the Minister of the Interior.
>
> At noon, the King, *Monsieur* the Dauphin, *Madame* the Dauphine, *Madame Duchesse* de Berry, and the Children of France, accompanied and followed by the entire court, went to the Orangerie, and Geoffroy Saint-Hilaire had the honor of presenting the gift of the Egyptian pasha to the King, along with a pamphlet on the giraffe that he had written, in which the history of this animal is described with care and precision.
>
> His Majesty wanted to see this singular quadruped walk and run; all the court's members were present, and her gait, especially while running, appeared absolutely extraordinary. The King questioned the learned academician for more than half an hour. His Majesty appeared very satisfied with his responses and deigned to show his complete satisfaction.
>
> At 3 p.m., the giraffe left for Paris, where she arrived safe and sound with her procession. A crowd of curious onlookers followed her right up to the King's Garden.

La Dernière Notice recounted that during the giraffe's presentation to the king, "a silk pouch suspended around her neck was opened, from which was pulled a parchment on which it was written in Arabic that the Giraffe, at the age of six months when she left Cairo, ate corn." Does that mean that the Koranic verse was on another parchment? The fact is that the announcement described not one, but "some amulets from Mecca." ... Whatever the case may be, on the many illustrations of the giraffe, she is sometimes represented with her talisman around her neck, at other times without it. Can we then date these drawings back to before or after July 9? Nothing is less certain. However, we can surmise that if one or more amulets disappeared from Zarafa's neck, the reason was precisely the status of these objects, that we would today qualify as *ostentatiously religious*. The very Catholic giraffe of the king could not be seen to be a Muslim!

This same *Notice* recorded another anecdote, fueling the anthropomorphism related to Zarafa:

> On these days, she displayed some very curious proof of her memory and intelligence: she picked out from the throng of spectators a few young Egyptians from

the school of Effendis established in Paris, their costume appeared attractive to her; she caressed them with a joy that without a doubt, contained melancholy. Their turban, their loose clothing reminded her of her first keepers, and the country of her birth to which she was still attached. In Africa, she would have fled these same men as enemies; in a foreign land, she greeted them with friendship, as if to push back the sorrows of exile.

A muse in spite of herself, the giraffe inspired a ballad by *Madame* the Countess of Oglou, *La Giraffe à Saint-Cloud,* set to music by Mr. Hardy. Sold for 1.50 francs in Paris, in the Petitbon music store, Bac Street, no. 31, *and through all music merchants* (figure 8.3):

Premier couplet
Voulez en vain dans mon âme at-tendri-e
Verser hé-las beaume [sic] *conso-la-teur*
Sombres chagrins ha-bi-teront mon coeur
Loin des miens, loin de ma patri-e (ter)
Deuxième couplet
N'est pas bonheur richesse qu'on envie!
N'est pas bonheur d'être admise à la Cour
Du Souverain, alors que nuit et jour,
On soupire pour sa patrie! (bis)
Troisième couplet
N'est pas bonheur loin de l'Éthiopie!
Et l'habitant de ces heureux déserts
Mieux aimerait chez lui porter des fers,
Que libre au loin de sa patrie! (bis)
Quatrième couplet
Bientôt je crois quitterai cette vie,
Empoisonnée au sein de la grandeur
Et par ma mort goûterai le bonheur
D'aller errer dans ma patrie! (bis)
Cinquième couplet
En attendant vous qui m'avez suivie:
Qui me quittez pour les rives du Nil:
Dîtes-leur bien, toujours qu'en mon exil
Ai soupiré pour ma patrie! (bis)

[*First couplet*
Your wish in vain within my tender soul to be

Figure 8.3. *Voyage de la Girafe, Grande Fantaisie Brillante* [The Giraffe's Journey, a Great and Brilliant Fantasy].
© Aynaud Collection.

To spill a balm consoling
Somber chagrin in my heart growing
Far from my own, far from my country (3x)
Second couplet
Happiness is not the riches of envy!
Nor happiness admittance to the Court we know
of the Sovereign, as night and day, to and fro
We go, longing for our own country! (2x)
Third couplet
Far from Ethiopia, and so unhappy!
And the residents of this good desert land
Would rather at home bear their swords in hand,
Than be free but far from their own country! (2x)
Fourth couplet
Soon this life I will have to leave,
A life poisoned in the midst of greatness
In my death I shall taste the happiness
Of wandering within my own country! (2x)
Fifth couplet
While waiting, you who have followed me,
Who leave me for the banks of the Nile:
Tell them well, that even in my state of exile
I have ever longed for my own country!]

July 12, 1827: After more than a month of silence, due without a doubt to his disappointment at not being able to embark at Châlons, as well as to his various health concerns, Geoffroy in the end wrote to the prefect:

> Finally, I have fully accomplished my mission: I received the giraffe from you, and I have delivered her to the king. The newspapers will have already informed you that it was last Monday that the giraffe was led to Saint-Cloud. The king had to come and see her the Thursday before in the Garden; it was thus agreed upon with his Minister: for orders for the troop's movements had been sent to the general staff in Paris; *Madame la Dauphine* saw it differently, she believed it would be more dignified if the king did not displace himself; she arrived two days early, and having questioned me, had it decided that what was possible should be done.

The style was charming. It was clear that the elderly, suffering naturalist, begrudgingly set the convoy back on the road but as for the question regarding the crown princess about the feasibility of the project, he could only give

CHAPTER 8

Figure 8.4. The Duchess of Angoulême.
Portrait of Marie-Thérèse Charlotte of France, Duchesse d'Angoulême, by Antoine-Jean Gros, 1816.
Palace of Versailles, Public Domain, France and the United States.

an opinion, finding no sufficient contrary argument. The king, urged by his daughter-in-law, the Duchess of Berry, would willingly have gone out to meet the giraffe, but that would have meant disregard for the watchful eye of the Duchess of Angoulême (figure 8.4).

> I was still suffering from urinary retention and from an inflammation of the urethra: I dragged myself along to Saint-Cloud, and I listened to my pains to discover what I must do, if I should advance or fall back. I took a lot upon myself: I appeared, and was able to satisfy the weighty audience that had fallen to me, which could have been shared by one of my colleagues, all present. But the king, to whom *Madame* the crown princess had recommended me, addressed himself only to me for a full hour, and took much pleasure in many of the details and points of view on the organization of species that I added to my responses. The king asked about how I had begun my mission; this gave me an opportunity to speak about *Monsieur* the Prefect of the Bouches-du-Rhône, and I did so with great feeling regarding the generosity with which I was honored in Marseille. The session, during which the giraffe allowed herself to produce all the niceties of which she was susceptible, ended with the sweetest of rewards for me. His Majesty wished to tell me that he had enjoyed all my responses, and he declared his full satisfaction.

Some interest was also displayed regarding the Egyptian handlers:

> The king wanted details on these men of service to the giraffe: I directed his attention toward Hassan who had already led a male giraffe to Constantinople, and to

Atir, a negro formerly enslaved by Mr. Drovetti. He went to the Minister of the Interior to order him to give two thousand francs to Hassan and one thousand francs to Atir that evening: this was carried out to the great pleasure of both of them.

The two groomsmen, Barthélémi and the younger negro Youssef, were thanked today by the administration of the King's Garden. The administration gave them more than sufficient means for their return. The first one made various speculations regarding the route and regarding Paris, and happily retired from service. He will say the opposite, as it is his habit to boast, he sets off declaring that he has done it all, and that he alone has the science and the capability to carry out the enterprise. Do you not recognize, *Monsieur* the Count, in this boasting some inevitable utterances from the land of the Marseillais?

No comment!

You will receive by post a sample of the brochure I wrote for the king, and of which I have only printed thirty copies. Nine artists have been engaged to reproduce the giraffe's image. I will once again take up my writing which touches on only a few points here, and I will treat the giraffe under all other related classifications. With my piece I will enclose another that is also of great interest, written by Mr. Mongès, presenting the whole history of the giraffe, with a note of all books in which she is cited. We are awaiting the engraving of the best drawing to be finished, and I will eagerly send all of this to you.

My Lord Count, as well as Madam Countess, please continue to keep me in your thoughts, and both of you, please accept the homage I owe you through my feelings of gratitude and respect.

On the same day, satirical expression continued with publication of the biting *Lettre de la Girafe au Pacha d'Égypte, pour lui rendre compte de son voyage à Saint-Cloud, et renvoyer les rognures de la censure de France au journal qui s'établit à Alexandrie en Afrique*. [Letter from the giraffe to the pasha of Egypt, in order to render an account of its voyage to Saint-Cloud, and to send French censored items to the newspaper starting up in Alexandria, in Africa]. Founder of the school in Athens, future academician, the Count Narcisse-Achille de Salvandy anonymously published these letters in the style of the *Lettres persanes* (*Persian Letters*) by Montesquieu:

Paris, from the Animal Palace or Royal Zoo, this Tuesday, July 10, 1827.
Good Prince,
My trip has ended. I am arriving from Court: it's a country that I already know by heart, for I see from afar and above, that is more quickly, and better than common observers....
This is what I find strange ... In the Sennar, the councils surround themselves with the most able, most active, most discreet, most renowned of the tribe. In

CHAPTER 8

Egypt, the most learned are also the most honored.... How can it be that in this France, whose culture is one of the most privileged in all the Universe, that shock awaits each talent, disgrace each illustration? It's an upside-down world here!

[The minister, wishing for turnover among the members of the committee censoring the press, asked the Giraffe to give him:] Her four negroes, who, neither knowing how to read or write, or speak French, would have marvelously suited the task of policing.... When this unhappy one had understood ... that he was threatened with the risk of abolishing national customs and open discourse, that he would have the task ... of strangling thoughts, of holding his scissors suspended over human reason, his poor head spun with it, he thought still that he had descended from my service to this foul trade, and he visibly weakened.... May the French obtain permission to subscribe to the Gazettes from our deserts to know what is happening in their own country! One has known for a long time that light comes from the Orient.

On August 8, 1827, the same author would publish the *Seconde lettre de la Girafe au pacha d'Égypte, en lui envoyant son album enrichi des dernières noirceurs* [Second letter of the giraffe to the pasha of Egypt, by sending to him his album enhanced with the latest black marks of censorship].

July 15, 1827: The petals offered by the king and appreciated by the giraffe spurred a rash of flowers: "Since one newspaper said that the animal loved roses, and savored their taste and scent with delight, all the curious onlookers going to see the beautiful quadruped in the King's Garden held out a bouquet of roses; the Fontenay fields won't be able to provide enough for everyone.... The rarest and most beautiful flowers were brought there, such as poppies or cornflowers gathered from the fields."[1]

And the zoo's director was obliged to insert a notice in the press to ask female visitors to "spare the giraffe some floral indigestion by being less generous in the sacrifice each one made to the foreigner."

July 18, 1827: "The giraffe is still the animal in fashion, and her historiographer, Mr. Ferlus, has just published the third edition of his notice on this astonishing animal that eats roses, drinks milk and procures for the stockholders of the botanical garden's bridge the extraordinary income of 650 francs a day."[2]

July 20, 1827: Le Figaro published a new schedule. The star had to be handled with care, "the giraffe had been seen each day since her arrival by thousands of people, and her preservation required some rest." *Le Figaro* being a satirical newspaper at that time, it announced the following week some "vols" (thefts or flights) in the King's Garden, but the penal code was not applicable in this instance: it was a matter of "vols" (flights) of birds.

All joking aside, *Le Gazette de Lyon* revealed on July 29:

> The crooks' headquarters is in the botanical garden. Over the past few days, a large number of handkerchiefs and watches have been stolen. The day before yesterday, in the *mêlée*, a Russian lord lost a superb snuffbox the sovereign of China had given to him. A lady in his company, soon found that her handbag had been cut open, and that the little golden snuffbox within it had disappeared.

Along with his museum colleagues, Geoffroy sent a portrait of Zarafa in a golden frame to the Egyptian pasha, as a sign of gratitude. In a letter,[3] he took sides regarding Egyptian liberation:

> I was told that a hippopotamus had been sent to Constantinople. May this tribute be the last that the viceroy sends to his Lord and Master. The interest of humanity and the prosperity of European nations ask that the Egyptian land be emancipated. Having a separate government and a government so usefully revolutionary as that of the Pasha, Egypt will achieve a great destiny, and the generous men who will have contributed to this memorable event will obtain praise and thanks from this age and for ages to come.

July 30, 1827: Geoffroy's words showed that he was up to date with international diplomacy. For, at the height of giraffomania, Drovetti came to Paris to bring an unofficial message from the pasha. Only three weeks after having met his giraffe, the king met with his consul–proof, if any were still needed, that Zarafa was "an instrument" of the pasha. The pasha, through Drovetti (whom one must consider his counselor), proposed to withdraw his troops from Greece on the condition that France would proclaim itself an ally, and help it to throw off the Turkish yoke: "Unable to count on the justice and benevolence of the Divan [Turk], he subsequently decided to direct his politics in the affair of the insurgents [Greeks] in such a way as to cooperate with Greek emancipation."[4]

Under pressure from the liberals, Charles X refused. The irony of the story is that France, a year earlier, had apparently made the same request through General Boyer, then among the pasha's general staff.... A French contingent would thus aid the Greek insurgents. The French, British, and Russians would join forces to crush the Egyptian fleet in Navarin.

August 2, 1827: Le Figaro made fun of Atir's success with women, using racially derived humor that appalls us today: "Two ladies tried to kidnap Mr. Atir, one of the giraffe's handlers. That would have been a really dark episode." And it went so far as to repeat the offense the following day: "The young Atir is learning music. He liked the black notes best."

CHAPTER 8

August 13, 1827: Let's remember that Muhammad Ali's lucky draw had given the king of England the most fragile of the young giraffes. This weaker giraffe, under the strong recommendations of the Consul Salt Henry, escorted by two Egyptian groomsmen and an interpreter, made the crossing to Malta, stayed there six months to avoid the harsh London winter, and finished their journey to England on board the vessel *Penelope*. On a boat whose deck had a wax canopy, she arrived in Lancaster on Saturday, August 11, 1827, at 6 p.m. She stayed in a warehouse until Monday morning at 5 a.m. Then a caravan escorted her to Windsor, where she arrived that evening. George IV, as well as his mistress, Lady Conyngham, were won over by the colossal animal (figure 8.5). He and his entourage would visit the giraffe more than once. English giraffomania had been launched, but just like the young giraffe, it would hardly have time to develop.

Geoffroy wrote in 1827 that the English giraffe, "it was said, probably perished in Malta." Had the naturalist maintained a grudge against those who had tried to steal the fruits of his research from him in Egypt at the start of the century? In any case, the English would suspect the French press of having launched a false rumor so as to be the only ones able to boast of possessing a giraffe. The fact is that the giraffe arrived in a pitiful state. It had wounded its feet during its initial trip on the camel's back, due to ropes that had been tied too tightly. An English technician had to invent a framework, with a belly strap, attached to a hoist, to keep the giraffe standing, and relieve pressure on its legs (figure 8.6).

As in France, painters were commissioned to immortalize the English giraffe: R. B. Davis represented her in different poses with an imaginary jungle in the background where palm trees were mixed with English oaks, and the Frenchman Jacques-Laurent Agasse made a painting and sketch of it. There again, caricaturists depicted that the king was besotted with the animal. Attributed to William Heath, *The Camelopard, or a New Hobby*, showed the king and his mistress, both with a paunch, sitting astride the giraffe. *The Great Joss and His Playthings*, painted by Robert Seymour, showed the giraffe snuggled up against a caricature of the king as an enormous big shot surrounded by his most costly possessions. In another by Heath, *The State of the Giraffe* shows the poor giraffe, dressed in stockings and ankle boots, suspended by its belly strap thanks to the hoist activated by the king and his mistress. In the margins, an added note: "I suppose that we'll have to pay for its stuffing afterward."

Finally, in *Le Mort, George IV, Caricature of the King Grieving the Death of the Giraffe*, the king, handkerchief in hand, grieves the dead giraffe, while his mistress is overcome by chagrin, and Lord Chamberlain is playing a funeral piece on the bagpipes. If we consider that this print appeared on August 11, 1829, the English giraffe died exactly two years after its arrival in England (figure 8.7).

Figure 8.5. King George IV and his mistress, Lady Conyngham, astride the English giraffe.
© Aynaud Collection.

Figure 8.6. The poor giraffe, no longer able to support the weight of its body, suspended via a hoist activated by the king and his mistress.
© Aynaud Collection.

Figure 8.7. The giraffe is grieved by the king and his mistress, to the sound of funereal bagpipes played by Lord Chamberlain.
© Aynaud Collection.

The English king would scarcely outlive the giraffe, as he passed away June 26, 1830. The skeleton and skin of the giraffe would be offered by the new king, William IV, to the Zoological Society of London. Its taxidermized remains would stay in the Zoo's museum until its closing in 1855, before being bought by a certain Dr. Crisp, zoologist. Thereafter all trace of it was lost.[5]

Back on August 13, 1827, on the same day in Paris, coming from Le Havre, rivals for Zarafa's popularity arrived: six American indigenous people, members of the Osage tribe. Received in the Terrasse Hotel on Rivoli Street, they became the latest attraction, following the trend launched by Chateaubriand with the publication of his epic Amerindian poem, *Les Natchez*. And in October, Geoffroy would write to the prefect: "The arrival of the Osage threatened to outdo us; but we have put up a good fight. Now at most they sit at their doorway to see these red-skinned men, but the giraffe continues to be sought after."

This most unlikely of comparisons between an Egyptian giraffe and an American indigenous tribe was the pretext for a charming piece: *Discours de la Girafe au chef des Six Osages (ou Indiens), prononcé le jour de leur visite au Jardin du Roi, traduit de l'arabe par Alibassan, interprète de la Girafe* [The giraffe's

CHAPTER 8

Figure 8.8. *Ah! Be our Great Mama . . . Beautiful giraffe.*
© Musée Carnavalet—Histoire de Paris.

speech to the chief of the Six Osages (or Indians), given the day of their visit to the King's Garden, translated from Arabic by Alibassan, interpreter for the giraffe]. The author was Honoré de Balzac, hardly disguised under his pseudonym, and who successfully evoked Zarafa's diverse avatars (figures 8.8, 8.9):

> My Lord Kihégasuhgah, I am very flattered by your visit; I have seen kings and princes; but I did not know of any in your color. Your costume has something that I like; this shaved head, those naked arms decorated with rings of silver or of white iron; that large wool covering that envelops you, it is all pleasing to my eyes. So there are such great persons everywhere, even as far as Missouri? . . .
> You wouldn't believe, my Lord, how mischievous the French are; there has been no critique that has not targeted my person. One said that my legs were too long and my tail too short; others admiring the beauty of my black pupils, seemed to mock my horns which they found too small relative to my body's mass; in the end, I would never finish if I were to tell you every joke they have told on my account. Also, please don't be angry if one sometimes makes jokes at your expense; the French mock everything.

Figure 8.9. What a program!
© Aynaud Collection.

CHAPTER 8

This was a means for the author to formulate a severe critique of the power in place:

> I heard that you had a sovereign you called Great-Chief. This Great-Chief, does he have the virtues of a great king? Does he like to hear the truth? Does he pay attention to the affairs of his subjects? Do his ministers support censorship, to avoid criticism of their administration? Does he send them away when they neglect the people's interests, in order to prioritize their own? Does your king find everything good, or does he sometimes include a judicious critique among his praises? Do the people often see him in person directing public works projects? Does he seek to uncover abuses, and does he hasten to eliminate them?

Certainly, the two attractions rivaled each other, as seen in this rude comparison between them:

> They have declared, I know, that your arrival in Paris has done a lot of harm to my reputation, to the popularity I hold; I assure you that this talk is baseless.... I continue to receive many visits, and Mr. Atir is here to support my progress. It is true that the public sees me for free, and is not obliged to bring money to the show in order to admire me in the Tivoli gardens or elsewhere.
>
> You must agree, My Lord, that many *savages* have come to Paris before you, without speaking of those who have not been.... Admit that your red skin cannot be compared to the beauty of my tiger-striped coat.... Do you have warriors of my size in your country? Do you have some large soldier who could not easily walk underneath my belly? Are the feathers that cover your head comparable to the straight horns that nature has planted upon my forehead? You are perhaps going to tell me that many Osages have heads endowed with such ornament: I wish to believe you; but they are doubtless of the same shape or have the same elegance.

In reality, according to the author, the Osages and the giraffe served the same cause: "One is even saying that you were led to Europe to occupy and distract people's minds from great political interests, and that the governments are paying for your trip. I admit to you, august Prince, that I don't believe a word of this."

The end is moving:

> I believe that all this talk is annoying you; I will finish; besides, you must be impatient to go and visit all the unfortunate companions with me in captivity; go and see the Lion, the Bear, the Monkeys, the Elephant; don't forget, please, a Giraffe that awaits you in the office of the Natural History museum. The scholars of this garden have stuffed it with straw to preserve it longer. This is the fate that awaits me!...

POLITICAL CARICATURE

Ah! my Lords, return quickly to your faraway climes, rejoice with a sweet liberty within your forests, on the banks of the Missouri or the Mississippi; leave the *civilized* peoples who may play some trick on you.... Go and smoke at the door of your lodging; go far from this garden, a veritable prison. I wish, with all my heart, that the great spirit might accompany you, that he will give you the grace to see your homes again; for me, I am spending my youth here without hope, and the approaching Winter already makes me tremble. *Adieu.*

Another competitor for attention who would scarcely know the honor of posterity was a whale, beached in Ostende. A lithograph made by Jacquemin and Langlumé portray it, reuniting near its sandy head every exotic star of the moment: the elephant, the Osages, some Chinese people, and Zarafa (figure 8.10).

September 4, 1827: The prefect responded late to Geoffroy's letter, but with strong homage for the naturalist:

> The many occupations I have had during the General Councils session have led me to declare belatedly how satisfied I have been with the details you have taken pains to write to me concerning the journey made by the Giraffe in your care, and on the circumstances of the arrival at her destination. I had no difficulty imagining the impression that this presentation must have made at Saint-Cloud, and I

Figure 8.10. The giraffe visits the whale.
© Aynaud Collection.

CHAPTER 8

congratulate you on having responded to the King's questions, since you have doubtless done so with all the expertise that distinguishes you. The Commission you have just completed in this Department is a fact I will remember through the details that have led me to appreciate your merits, and to foster relationships that will always be of interest to me.

The follow up, as a P.S., was not written by his secretary but in the prefect's own hand, indicating his emotion in this manner:

P.S. I wanted to tell you in my own hand that I still conserve a precious memory of the few moments you spent among us, and I assure you that nothing of this will fade from my mind or my heart. It was a great pleasure to learn from you that our pupil had been so well received after her arrival, and thanks to your care, in perfect health. Upon seeing her, think of her tutor from Marseille, and never doubt the sentiments that you have inspired in me. Your devoted Servant and friend.

To Zarafa's credit, this strong friendship was launched between two men who knew how to work together!

October 20, 1827: Under England's influence, the tripartite alliance with France and Russia had as its goal to impose mediation in the Turkish-Greek conflict, to obtain an armistice as well as Greek autonomy. But the battle of Navarin provoked the destruction of the Turkish-Egyptian fleet. England, considering this victory a deplorable misunderstanding, pressed Turkey to concede, but the proud Mahmoud was too stubborn.

Seeking to play his cards right in this dangerous game, Muhammad Ali recalled his troops and asked the sultan to give him Syria, as recompense for the loss of his navy. The sultan refused, and in 1831, the pasha invaded Syria, with the help of Soliman Pasha, the Turkish name of the French colonel Sève. As he claimed victory after victory, the pasha would finally force the sultan to concede. He would in the end receive all of Syria, Palestine, and the Adana region, in the south of present-day Turkey.

October 21, 1827: The French government signed a contract of sale to acquire the second collection of antiquities from Consul Drovetti. Destined for the Louvre museum, it was bought for 150,000 francs, payable in five installments over the period ending in October 1830.

October 22, 1827: Geoffroy sent the prefect a letter that would not be posted but rather given to a special messenger, Hassan, returning to his homeland: "Still one more note about the giraffe: you have allowed me to speak to you of this highness. Your good will for her will save me from any ill will due to having inconvenienced you."

Figure 8.11. In her stable at the Rotunda.
© Aynaud Collection.

Zarafa was lodged in the oldest but most beautiful of the zoo's buildings, the rotunda. Built in 1805 according to plans copied from the pentagonal cross of the Legion of Honor instituted by Napoleon, it housed exotic animals, first from the palace of Versailles, survivors of the revolutionary upheaval (figure 8.11).

> She has just moved to her Winter quarters. They comprise one hexagon of the Zoo's large building called "the Rotunda," appropriated for this use. The doors

CHAPTER 8

were joined; one will open and the other will remain closed depending on the season and the access she will need to the inside or the outside of the building. A wooden floor was placed on the pavement. The lodging's walls were covered with straw mats, elegantly matched. Apart from the fact that the room is large, it's truly the boudoir of a young mistress. To add to the charming effect, the negro Atir goes up to his bed via two ladders, because of a large landing between them. The two of them see each other at eye level in the high space of the lodge: the giraffe lifts her head and Atir climbs to elevate his head as well.

The antechambers of the giraffe's room are heated by stoves, but as this dry heat only partially works with the phenomenon of respiration, other more efficient stoves are placed in her stall, which are the companions given to her, her cows, and other Indian cows gathered there. In this way, we reach six degrees higher than the regular temperature. When the cold becomes harsher, we will double and triple the number of animals.

All taxes have been counted, from the post office and the Pont d'Austerlitz, and what the giraffe has procured in income growth, by the movement she has started within the population, and taking into account the extraordinary money spent by this population on various consumable items. In July, it reached 60,000 [francs], and in August, 40,000. The Pont d'Austerlitz received 8,000 more than the previous July 1826; and last August, 5,650 more than the previous August.

Hassan is leaving us, and I am entrusting him with this letter for you, My Lord Count. The king has given him a gift of 2,000 francs. We have treated him with respect, and he has merited our care for he has remained well and truly, loyal to his animal. He is leaving in rather poor physical shape, with a hernia, one bad eye, and I think a chronic liver ailment. God will take pity on him one way or another. We gave him one franc per league [about four km] for his return as far as Toulon, or 226 francs. I flattered him, My Lord Count, that you would continue the generosity that you have deigned to give him in the past.

Atir remains with us. He is a real French knight of good fortune. He is talked about; for *Madame* la Duchesse de Berry wanted me to confide in her some of his adventures or misadventures. With princesses, it goes in one ear but quickly goes out the other.

Without a doubt, it is understood that certain worldly French Parisian women were not insensitive to the exotic charm of the groomsman, and his escapades fed the local gossip of the capital.

The prefect's personal note had touched Geoffroy:

I had promised your librarian that I would send along for his library two of my brochures that were being printed in Paris when I was in Marseille. I take the liberty of enclosing them.

My Lord Count, please receive my thanks for your last letter's note *in your own hand*: I was really touched. Please receive along with my thanks the homage of my very respectful and devoted sentiments.

November 24, 1827: After Villèle's dissolution of the Chamber, the legislative elections gave the left 170 representatives, versus 125 to the government in place, provoking the fall of the reactionary minister. During the harsh winter, doctors named the nasty influenza, flu "à la giraffe," so that to ask about the health of a third party, one asked, "How's his giraffe?" and at times, one would hear the response, "He died, his giraffe was really severe."[6]

August 7, 1828: A giraffe arrived in Vienna, after having navigated from Alexandria to Venice and crossed the passes separating the Adriatic Sea from the Danube valley. To cross the Alps, the animal's hooves were covered in custom-sized leather shoes. The same giraffomania was launched on the spot. The Schönbrunn Zoo must have encountered a similar crowd to that seen in the French zoo during the previous year. The Austrian giraffe would inspire a new dance, the *Giraffe Gallop*. At the entrance to a ball given in the animal's honor, young women received a bouquet of flowers topped with a sugar giraffe head. The Viennese porcelain factory created an inkwell decorated with giraffe silhouettes. But the Austrian giraffe would die eight months after its arrival, following its exhausting journey.

In France, a female elephant from India joined the only pachyderm of the King's Garden. The public was enthused about the new couple. But they would wait in vain for this arranged marriage to bear fruit.

August 17, 1828: Following French intervention in Morea, the Egyptians of Ibrahim Pasha, allies of the Turks, re-embarked. Three days later, after two armies marched on Constantinople, the Russians took Adrianople and forced the Turks to negotiate.

1829: The "stereotyped *Almanach*" (titled "Le Messager boîteux"), in 1829 was called *"à la giraffe,"* and explained that the animal was still "the emblem of novelty in France." Charles X asked for a study of his predecessor's project—to bring to France an antique obelisk. Joseph de Laborde and Champollion then began in Egypt what we would call today a feasibility study. In a letter from July 4, the great Egyptologist told of his preference for the obelisks from Luxor, especially because of their better state of preservation.

The affair in Algeria worsened. Under the government headed by Martignac, the French fleet tried to organize the blockade of the port of Algiers, which turned out to be an ineffectual and somewhat ludicrous move. France attempted to negotiate, but the dey, encouraged by the perfidious Albion (England),

CHAPTER 8

rejected his propositions. On August 3, 1829, even a parliamentary vessel, *La Provence*, sustained cannon fire upon leaving the port of Algiers. Things seemed to be at an impasse when, on August 8, 1829, King Charles X, of his own accord, replaced the president of the ministry with Prince Jules de Polignac, an ultraroyalist who was very unpopular in public opinion, that is to say, unpopular with the sixty thousand bourgeois holding the right to vote.

September 14, 1829: The Treaty of Adrianople put an end to the war for Greek independence, without precisely establishing the borders or the status of the country as vassal or independent. Civil war would follow the struggle against the Turks, and the heroes of the emancipation would become divided.

An *Invocation à la Giraffe*, words and music by "F. G.," illustrated with a lithograph by Langlumé, already contained a biting perspective:

Chœur
Sur notre terre hospitalière
Viens, fille des déserts brûlants
Et par ton élégance altière
Giraffe, giraffe, inspire nos accents
Douce giraffe, inspire mes accents
Grande giraffe, inspire mes accents
Giraffe, inspire mes chants.

Premier couplet
Quand j'aperçois tant de gens d'importance
Cerveaux étroits marchant le front bien haut
Je vois que nous avions en France
Mainte girafe comme il faut! (bis)
Gens orgueilleux qu'on ne saurait atteindre
Lorsqu'on vous parle, hélas, qu'on est à plaindre!
Dès qu'il s'agit de s'entendre avec eux,
D'être giraffe on serait trop heureux. (bis)
Deuxième couplet
On ne voit pas chaque bête envieuse
Déshonorer le paisible jardin
Et la Giraffe vit heureuse
Entre l'Éléphant et Martin (bis)
L'envie, hélas! chez nous si meurtrière
Frappe l'émule entré dans la carrière;
Pour rencontrer des rivaux généreux

D'être giraffe on serait trop heureux. (bis)
Troisième couplet
Pour éviter la figure importune
Du vil flatteur qui s'agite en rampant.
Pour courir après la fortune
Et voir de loin l'homme puissant: (bis)
Enfin, pour voir, Mesdames, au spectacle,
Où vos chapeaux sont un cruel obstacle,
Chacun allonge un col ambitieu;
D'être giraffe on serait trop heureux. (bis)
Quatrième couplet
Privé d'accès au temple de mémoire,
Nos beaux esprits vont maudissant le sort;
La Giraffe ignorant sa gloire
S'immortalise sans effort: (bis)
Tout reproduit son image fidèle,
Robes, bijoux, broderie et dentelle;
Quand sans génie on veut être fameux
D'être giraffe on serait trop heureux. (bis)

[*Chorus*
To our hospitable lands
Come, daughter of burning desert sands
And with your high elegance
Giraffe, giraffe, inspire our accents
Sweet giraffe, inspire my accents
Tall giraffe, inspire my accents
Oh Giraffe, inspire my chants.
First couplet
When I see so many people of importance
Narrow skulls on heads held high indeed
I see that we have in France
As many giraffes as we need! (2x)
Proud people, much too high to attain
When you tell them, alas, that you wish to complain!
When to talk with them one finds an opportunity,
A giraffe one would only too happily be. (2x)
Second couplet
One does not see every animal enviously

CHAPTER 8

Dishonor the peaceful garden
And the Giraffe lived so happily
Between Elephant and bear Martin (2x)
Envy, alas! inside us such insidious fare
Strikes the disciple entering his career;
To encounter such generous rivalry,
A giraffe one would only too happily be.
Third couplet
To avoid the unfortunate face
Of vile flatterers, writhing as they grovel.
To follow fortune's pace
And spy from afar those men so powerful: (2 x)
Finally, to see, *Mesdames*, at the spectacle,
Where your hat is a trying obstacle,
Each man his ambitious collar so lengthy;
A giraffe one would only too happily be.
Fourth couplet
Deprived of access to the temple of memory,
Our beautiful intellects leave, their fate cursing;
The Giraffe, ignorant of her glory
Is immortalized without even trying: (2x)
All things recall her loyal face,
Dresses, jewelry, embroidery, lace;
When without genius one wants to live famously
A giraffe one would only too happily be.]

Upon the death of the prefect Christophe de Villeneuve-Bargemon, Zarafa's story took another turn, conferring a new message.

9

ZARAFA'S LEGACY
An Existential Fable

In his *Épître à la Girafe* [Epistle to the giraffe], Charles D.... had already expressed a note of nostalgia, foreseeing the public's fleeting infatuation with the giraffe (figure 9.1):

Tu perds déjà cette joie éphémère
Dont les premiers honneurs avaient su t'enivrer
Tu penses sans doute à ta mère,
Tu la vois au loin s'égarer
Tu la suis sous le frais ombrage,
Vers la fontaine du vallon....
Si tu pouvais redevenir sauvage,
Et près d'elle arriver d'un bond!
Eh, que te font à toi ces huttes opulentes,
Ces esclaves, ce soin, cette paille apprêtée?
Tu devais vivre pauvre en tes plaines brûlantes,
Mais pauvre avec la liberté!

[Already, you're losing that joy so fleeting
Whose first honors had intoxicated your way
Surely you think of your mother retreating,
Off in the distance, lost and far away
In the cool shade you follow
Her toward the spring in the vale....
If only you were wild again tomorrow,
And could leap to her side in the swale!

CHAPTER 9

Figure 9.1. Giraffe at two years of age.
© Musée intercommunal of Étampes.

What do these opulent huts mean to you at present,
These slaves, all this care, this straw piled high?
To live poorly on scorching hot plains you were meant,
To be poor, and yet free under the sky!]

March 3, 1830: In a speech from the throne, it was the first time Charles X raised the idea of a punitive expedition to seek debt reparation as well as to destroy the hideout of corsairs in the Algiers regency and put an end to slavery.... Confronted by the revolt of 221 deputies, the king quickly needed to improve his image. Count Louis de Bourmont, minister of war in the Polignac government, was appointed commander in chief of the African expedition. "Monsieur de Bourmont wanted to be Marshall, and he deserves to receive the staff!" wrote *Le Figaro*.

The English didn't hide their disapproval, but nothing was to be done about it. On the 25th of May, 453 ships set sail from Toulon, with 83 siege engines, 27,000 sailors, and 37,000 soldiers. On June 14, the French troops disembarked on the beach at Sidi Ferruch, twenty-five kilometers from Algiers. From then on called "zouaves," these soldiers wore the fashionable, ample trousers inspired by Zarafa's official groomsman, Atir.

The fleet bombarded the city's defenses. The dey capitulated on July 5, after several days of difficult combat against the Turkish troops, who killed 415, and wounded 2,160 men of the expeditionary forces. Forty-eight million francs from the treasury were enough to cover the expedition's costs. The French soldiers set about shamelessly ransacking the city. The occupation of Algiers was greeted with indifference by the French. King Charles X would be chased from the throne a few weeks later.

June 30, 1830: The end of "giraffe fashion" appeared to coincide with the declining popularity of Charles X among his subjects. This parallel did not escape French author Honoré de Balzac, who wrote these prophetic lines several weeks before the Revolution of 1830 for the illustrated newspaper, *La Silhouette*:

> The giraffe in her current position is a great moral idea, an eloquent philosophical teaching.... What name had more impact? What animal has ever been so popular? But what does Solomon say? Vanity of vanities, all is vanity.... And yet the giraffe is an eloquent summary of what wise men and all philosophers have said most strongly about oblivion of the famous, she is living proof of it; for today she is rejected, forgotten, no longer visited except by the backward provincial, the idle nanny, and the simpleton or the naive. Many men should learn from this striking lesson, and foresee the fate that awaits them.

And Balzac clarified the analogy further: "Thus, consider today the president of the Council of Ministers: for the past several months all have been talking about him as well; the giraffe had not generated more subjects of conversation, eloquent declamations, spiritual epigrams than he; yet after a few days, he too will be forgotten just like the giraffe."

July 26, 1830: Charles X announced *five infamous rulings*: suspension of freedom of the press, new dissolution of the Chamber, modification of the

CHAPTER 9

electoral process, the convocation of electors in September, and the nomination of royalist high public servants. Inspired by Thiers, some journalists of the opposition protested. In three days, this second French Revolution caught fire and saw the erection of barricades in Paris, the taking of the Louvre and the Hôtel-de-Ville. As a secondary result of the riots, gates to the Botanical Gardens were closed, and all visits to Zarafa were forbidden.

After withdrawal of the rulings, deputies of the liberal opposition confiscated this popular second revolution to the benefit of the Orléans monarchy, appointing then Louis-Philippe as lieutenant general of the kingdom. On August 2, Charles X abdicated the throne, and on the 9th, Louis-Philippe I became king of France. Geoffroy applauded the events of 1830 but, mustering the courage that recalled his youthful exploits, he saved the day for the archbishop of Paris, Monseigneur de Quêlen, offering him asylum in the museum.

The king and his intransigent daughter-in-law, the Duchess of Angoulême, were forced into exile once again. From the point of view of the emigrant Zarafa, one could say that justice had been served. A partisan of the Revolution of 1830, Stendhal would, however, write with nostalgia, "It may take centuries for the majority of European peoples to attain the degree of happiness they held under the reign of Charles X."

October 9, 1831: After the assassination of Capo d'Istria, at the head of the temporary government of Greece, a commission called to the throne a seventeen-year-old Bavarian prince named Otto, a dedicated Philhellenist. The independence of Greece was fully recognized, and its northern borders extended from the Gulf of Arta to the Gulf of Volo.

November 21, 1831: The Canut revolts in Lyon soon followed. Those who wove "giraffe fabrics," which were all the rage in high society, no longer accepted the economic insecurity associated with their trade. The silk workers who were making eighteen to twenty-five cents (*sous*) per day instead of the four to six francs they had earned under the empire, took over the city. On December 5, Marshall Soult brutally reestablished order in Lyon.

December 22, 1833: The obelisk from Luxor arrived in Paris. The twenty-five-meter high column of rose granite was consecrated the "second vertical marvel in Paris," after Zarafa.

June 28, 1834: Kings moved on; Zarafa stayed. Satirical commentary using the giraffe was not yet exhausted. The engraving, *Ménagerie Royale* [Royal Zoo] in *Le Charivari*, unsigned and attributed to Daumier, is a caricature of Louis-Philippe depicted as a large elephant near a giraffe in military uniform, perhaps one of his sons. Referring to Zarafa and her elephant companion, the caption said only, "Portrait of two dining companions well-known in the king's garden" (figure 9.2).

Figure 9.2. King Louis-Philippe depicted as a giraffe.
© Aynaud Collection.

1835: Someone had the idea of marrying off Zarafa, but the male destined for her would never leave Italy. "Italy currently possesses an apparently male giraffe, and may we not give up hope of soon seeing him in the Paris zoo with the individual that nearly all of France has already seen."[1]

Zarafa would remain a spinster.

1836: Disconcerted by the loss of their giraffe, the English sought to obtain from Muhammad Ali a couple of male and female giraffes through diplomatic channels; their wish came true that year.

Two years after its arrival in the capital, the Luxor obelisk was finally set in the center of the *Place de la Concorde*, making it one of the most beautiful squares in the world. Credit goes to Charles X, who in 1830 had appointed Baron Taylor royal commissioner to the Egyptian pasha, responsible for negotiations. November 6, 1836, the deposed king died of cholera in Goritz.

1838: After ten years of good and loyal service to the giraffe, Atir returned to Egypt. Witnessing him brush Zarafa's long mane every day, someone created the long-lived expression: *"Faire ça ou peigner la girafe!"* [Might as well brush the giraffe again!]

March 18, 1838: England would obtain its revenge with regard to its first giraffe's destiny. Several days before spring 1838, two giraffes from the London

CHAPTER 9

zoo, the male Guib-Allah and the female Zaida, mated for the first time, and then again on April 1.[2] On June 13, after a pregnancy of 444 days, the female gave birth to a male calf:

> After one minute, he took his first breath, accompanied by a spasmodic shaking of his whole body; he assumed a comfortable pose, continued to breathe in a very regular manner and a half hour later, made efforts to stand up. First, he put his knees forward, then he soon began to walk, though vacillating a bit; he circled round his mother. She did not welcome him at all as one had hoped; all that could be obtained from her was a look of astonishment at the young, tiresome one who continued to be a complete stranger to her; so he quickly fell ill and on June 28, he died.[3]

Only later would they understand this maternal indifference: someone had lent a hand to assist with the birth. A second fertilization occurred on March 26, 1840, and 431 days later, Zaïda again gave birth to a male calf: "The mother, left alone during labor, showed all the tenderness expected of her toward the little one; the young animal was soon full of energy, he continued to live, and was sent on later to the zoological garden in Dublin. After one week, he was already six feet tall and at three weeks of age, he was eating the same diet as his mother, and chewed his cud with the same ease."

After that, Zaïda gave birth four more times, in 1844, 1846, 1849, and 1853.

1839: In order to break the solitude of the former star of the King's Garden, a second giraffe was sent to Paris by the doctor Clot-Bey. This eminent Frenchman from Egypt took advantage of Zarafa's experience to ensure the best conditions for transporting his protégée: he recruited a specialized caretaker and handler, arranged transport on the Nile River in a felucca stocked with straw and covered with a tent, and she crossed the Mediterranean in a deep cargo hold kept dry with straw hatching, with a custom-made canvas coat for Egypt and one made of wool for France.[4]

Zarafa would thus benefit from the company of an equal.

1840: From the sultan of Constantinople, Muhammad Ali obtained the governance of Egypt with hereditary rights, founding a dynasty that would reign until 1853.

Having gone blind, Geoffroy offered his resignation to the chairs of the museum and the Sorbonne. But they insisted that he remain tenured at the university until his death. His son would follow in his footsteps. Isidore Geoffroy Saint-Hilaire, who had shown a very early aptitude for mathematics, in the end would pursue a career in natural history and medicine. At the age of nineteen, he became his father's assistant naturalist. Between 1832 and 1837, he published his main work on teratology, *Histoire générale et particulière des anomalies de l'organisation chez l'homme et les animaux* [General and specific history of anomalies in the organization of man and animals]. In 1829 he began to teach his father's courses on ornithology, and during the next three years, he taught zoology and teratology in the *École pratique*. He became a member of the *Académie des Sciences* in 1833. In 1837 he taught in the Paris *Collège des Sciences* and, the following year, went to Bordeaux to set up a similar college. In turn, he became inspector at the Paris *Académie*, then professor at the *Muséum national d'Histoire naturelle*, after the departure of his father.

In December the ashes of Napoleon (the one who had given Geoffroy the opportunity to encounter Egypt), were returned from St. Helena and transferred to Les Invalides. By organizing this display, King Louis-Philippe I tried to gain the respect of a growing French Bonapartist contingent.

1842: Frédéric Cuvier, George's younger brother, gave news of Zarafa in his text, *Histoire naturelle des mammifères* [Natural History of Mammals], written in collaboration with Geoffroy:

> When Buffon wrote the history of the giraffe for the first time, he had only at his disposal some descriptions that mixed errors with facts offered by the ancients and a few modern writers....
>
> I will content myself to take up the history of this animal where Buffon had left off, and add observations regarding this species around which favorable circumstances today have encouraged a proliferation of naturalists. Not only had it not been viewed alive by Christians for more than three hundred years, but even the remains of the animal were rarely seen in the beginning of this century, while more recently, close relations established with Egypt have enriched several of the large zoos of Europe with live animals....
>
> For the past sixteen years, the Museum's Zoo has been keeping the female whose portrait we are providing.... We have seen ours gallop along the paths of the Botanical Gardens, in moments of gaiety, leaving the guardians leading her well behind by several bounds.
>
> Our Giraffe easily and happily lowers her head to the height of the people who approach her: ours has never uttered a single cry.... Our female, who was eleven and a half feet tall upon her arrival, has grown by two feet since then, and is now thirteen and a half feet in height; but she stopped growing seven years ago, and consequently, ceased to develop further.... It is notable that the stuffed animal of the Museum office, brought from Cape town by Levaillant, is much taller than our live Giraffe.

In the beginning of the year, a second giraffe arrived from Abyssinia but then died in Toulouse. She was described in this way: "Magnificent in ap-

CHAPTER 9

pearance, with bizarre contours, singular in her gait, colossal by her size, and inoffensive in character."[5]

The observation was refined and allowed for revelation of astonishing anatomical details:

> Consider this beautiful ruminant close up; examine its head, with such a slender profile, its very gentle gaze; see its nostrils, which by an admirable mechanism, can close by the animal's wishes when a cloud of sand raised by the simoon [desert wind] might wound the very delicate membrane covering it; examine especially the movements of its eyes, which, protected by such a prominent eye socket, can perceive the enemy from far off, from the front, the side and from behind; contemplate finally this third, half-transparent eyelid which lowers over the eye like a curtain, and protects this organ from the harshness of a too bright sun or from burning sand.

The sociologist Charles Fourier, inventor of the phalanstery, found in the giraffe inspiration for his philosophical analogies that were sometimes difficult to follow. And his historical references were somewhat erroneous, notably when he confused Algeria with Egypt:

> And also, what do you think about what the celebrated traveler Levaillant had brought to us; what gift do you think the Dey of Algiers gave in the past to Charles X, the first one returning with a taxidermized giraffe, the other sending a live giraffe to the king's garden? France has naturalized this hieroglyphic animal of truth. This animal, as everyone knows, raises its forehead above all others. That is the propensity of truth, to rise above errors.
>
> The giraffe, as an old author has said, is very beautiful, gentle and pleasant to look at. Truth is also very beautiful; but the giraffe should not be used for work, as she would not know how to adapt to our ways. God has thus reduced her to uselessness by giving her an irregular gait that agitates and strains any burden set upon her. And so, we prefer to leave her inactive as, among us, we also exclude from tasks the truthful man whose character would disturb all traditional customs and acts of will. Truth among us is beautiful only if left inactive, and the giraffe by analogy is only admired when she is at rest; but in her walk she excites cries of derision just as truth excites cries of derision when it is active....
>
> It is much worse in politics, where truth flourishes even less, and in order to represent this pressure that is truth, this invincible obstacle to its development, God has cut down the giraffe's horns at their base; they may only sprout but not spread their branches. God's scissors have cut them at their base, just as the scissors of authority and public opinion among us have cut truth off when it appears and forbid any growth.... One can see in this explanation that God had not created anything without its use, including the giraffe, which is of perfect uselessness.
>
> But what purpose does the giraffe serve?

Geoffroy asked himself this question in 1827, right after his mission:

These are the impenetrable designs of Providence. She is for black Africans what the wild animals of our forests are for Europeans. People say that the deer that populate them, embellish them, and enliven our countryside provide relaxation and pleasure for the great ones of the earth. Why would one not say the same of the giraffe? There is a perfect analogy between one and the other, except that the woods are the places of refuge for our wild animals, while deserts are inhabited by giraffes and antelopes. It is certainly of no use to explain how and why the nature of things has decided this.

June 19, 1844: Étienne Geoffroy de Saint-Hilaire would die surrounded by those close to him as well as by unknown students who insisted upon watching over his final moments. His death would signal the disappearance of the last professor appointed at the creation of the museum, one of the predecessors of evolutionism, and representative of a school that the Enlightenment and the Revolution had elevated to the top of the world in its class.

Geoffroy had never published a single clear and precise text aiming to show *transformism*. However, he had collected much data in favor of evolution: unity in organizational plan in vertebrates, and demonstration of the futility of the preexistence of germs, which was the theoretical basis of the fixism of species. Lamarck had already grasped the main points, in particular the ramifying series, but he had scarcely provided any concrete, solidly founded arguments. On the other hand, Geoffroy had amassed some elements of proof, even if he had not understood the mysteries of evolution as well as his elder.

Persuaded to be on the right track after more than twenty years of work and occasional struggle in favor of transformism, did Geoffroy have enough resources to undertake a rather advanced work of synthesis? We saw him concede to discouragement in 1799, after Bonaparte's departure; perhaps he had simply renounced his study out of lassitude. A moment arrives when it is tempting to pass the torch on to others, when too much accumulated dogmatism, stupidity, and ignorance get in the way. That torch would be taken up by Darwin in 1859.

The second controversy that had opposed Geoffroy and Cuvier corresponded to a unifying plan of the animal kingdom. Geoffroy demonstrated that the animals were constructed according to a double axis, antero-posterior, but also dorso-ventral, inverted in insects, crustaceans, and cephalopods, versus that of vertebrates. Cuvier had refuted his argumentation, but in 1996 American geneticists proved that Étienne Geoffroy Saint-Hilaire had been right.

January 12, 1845: Six months after the passing of this great man, Zarafa, the first giraffe in France, died at twenty-one years of age. Knowing today that

CHAPTER 9

the life-span of a giraffe in the wild is from fifteen to twenty years, and at least twenty-five in captivity, Zarafa had died at an honorable age, which proves that she had received quality care. The catalogue of August 1840 to December 1845 of the comparative anatomy laboratory of the National Museum of Natural History noted the reasons for her death: "Death from the effects of a phthisic tuberculosis of both lungs and especially the left side, which was linked to the rib pleura of this same side." Mr. de Blainville made several drawings for the Museum's records.

The master taxidermist Poortmann and his team inaugurated a new technique for the occasion: "They started by drawing her exact profile on the wall of their workshop and then had some sturdy pieces of wood cut, according to this design, from which a life-size statue of the animal was sculpted. Then the only thing that remained was to cover this masterpiece with the skin which assistants had carefully prepared in the meantime."[6] Right up to her last avatar, Zarafa benefited from the utmost respect and would thus decorate the great gallery.

Like all great figures, Zarafa left a taste for the irrational in her wake. This is what *Le Caducée de Marseille* reported, thirty-four days after the death of the giraffe: "At that time, a small cabaret, established in the street Rifle-Rafle, next to the Portalet, had taken for its name *La Giraffe*, and had the animal's image painted on the wall. A singular coincidence! It was said that the day that the owner of the cabaret white-washed the image of the giraffe, she died in Paris."

But the great lesson the nineteenth century took from the saga of Zarafa is existential: "Look, giraffes of the moment: this is how glory passes! Contemplate this skin stuffed with straw that the caretakers dust every morning, and consider that none among you after death will leave this much of a trace!"[7]

August 2, 1849: Muhammad Ali died in Cairo. With the death of Drovetti three years later in Turin, the last actors in the fabulous story of Zarafa would disappear.

But would her image be erased so easily? On April 14, 1850, *Le Charivari* published three new caricatures, associating this animal's legend with some indicators of industrial progress.

In 1859, exactly one-half century after the publication of the Lamarckian theory of *transformism*, Charles Darwin developed his theory of *natural selection*, taking into account the example of the giraffe. Giraffe herds include individuals with necks of varying lengths. Because they could graze upon the highest leaves, giraffes with the longest necks survived and reproduced better. Gradually, the giraffe population evolved toward individuals with the longest neck: "The high stature of the giraffe, the lengthening of its neck, of its forelegs, of its head and its tongue created an animal admirably adapted to graze upon the high branches of trees. It can thus find food placed outside the reach of other ungulates living in the same country; which during times of scarcity, procured great advantages for it."

Ten years later, the naturalized remains of giraffes became commonplace. With the opening of the Suez Canal, inaugurated on November 17, 1869, a significant number of live giraffes were brought to Europe and America. A year later, a merchant rented a veritable Noah's Ark on the Kenyan coast, including no less than 14 giraffes among the 130 animals on board.[8]

Zarafa's remains were no longer an attraction. But she would make a comeback in contemporary France in a manner as fantastic as her arrival in the country.

During the First World War, she was said to be hosted by the Verdun museum. In this martyred city, pounded by bombing, her naturalized body had remained intact on the second floor of the bishop's residence, defying the surrounding human horror. But who had gotten the absurd idea of bringing the animal to the front line to impress the assailants? And what could the enemy have thought, seeing this horned head with its benevolent gaze rising from the trenches? It was not a Trojan giraffe, but perhaps a new form of periscope.... By the end of the first volley of cannon fire, she had been removed and had disappeared. Rest assured, a posterior inquiry would prove that this giraffe, the one who had been honored with an official reception upon her arrival on the banks of the Meuse, was not the one given to Charles X. No, Zarafa had not died a second time in the trenches of Verdun!

1931: Dr. Étienne Loppé, curator at the Museum of Natural History in La Rochelle, had the honor and advantage of being able to dig into the collections of the national museum. His choice awarded a new home to Zarafa's remains:

> At the beginning of last July, Professor Bourdelle, of the National Museum of Natural History, wished to offer me a certain number of large mammals for the collections of the Lafaille museum, taken from *duplicatas* within our great National Collection.
>
> I therefore chose an important series of ungulates, pachyderms and ruminants that our collection does not possess.
>
> This gift included: two tapirs, two elephants (Indian and African), two hippopotami, one rhinoceros, one giraffe, one chital, several large antelopes, one moose, and one wild ox.
>
> Some of these animals came from the research of great traveling naturalists of the nineteenth century, Diard and Duvaucel, J. Verreaux, and Delessert, and had been studied by Cuvier himself.
>
> The rhinoceros from Sumatra and the hippopotamus from Liberia are two rarities of the first order and the giraffe, a historical piece. As the first live one seen in

CHAPTER 9

France, offered to Charles X, by the pasha of Egypt, she garnered considerable interest during the Restoration.⁹

Whether by coincidence or as an explanation for shedding stock, an international colonial exposition in the *Bois de Vincennes* took place from May 15 to November 15, 1931. Inaugurated by the French minister of the colonies, Paul Reynaud, and the president of the republic, Gaston Doumergue, this display marked the apotheosis of the third republic, in order to convince public opinion of the legitimacy of the conquests. The *commissaire général* of the exhibit was Marshall Lyautey, who, at age seventy-seven, thereby accomplished his last commissioned project. The countries addressed covered a good part of Sub-Saharan Africa, Madagascar, North Africa, Indochina, Syria, and Lebanon. The expo, which occupied two hundred hectares in the *Bois de Vincennes*, numbered two thousand exhibitors. It was a huge success, hosting several million visitors.

Doctor Loppé buckled down to a long inquiry, at the end of which he responded to his colleague at the Verdun museum that La Rochelle would house Zarafa from then on. In fact, Professor Alfred Lacroix owned the giraffe "said to be Daubenton's giraffe," which was taxidermized at the museum of Paris by Delalande. The true Zarafa took her place, then, near this port with two towers, beside Napoleon's dromedary, that *vessel* the emperor had taken to explore the Suez desert during his Egyptian campaign, which was later replaced by the orangutan of Joséphine de Beauharnais.¹⁰

September 25, 1950: More than a century after her passing, Zarafa was reborn thanks to the curiosity of one man. Gabriel Dardaud, director of the *Agence France-Presse* for Egypt, wrote to the curator of the museum in La Rochelle, which became the Muséum Fleuriau:

> For over a year, I have been doing research for the Egyptian Institute on the giraffe offered by the Egyptian dynasty's founder to the king of France. I have uncovered an abundant amount of documentation on the subject in the archives in Cairo, in those of the Quai d'Orsay and the department of the Bouches-du-Rhône as well as in the National Library of Paris. However, I thought that the giraffe herself had disappeared in the bombing of Verdun and I was very surprised to learn that she was preserved in the museum in La Rochelle.¹¹

It was the story of a male orphan moved by the destiny of a female orphan. A pupil at the *Orphelins d'Auteuil* (Orphans of Auteuil), Dardaud had obtained a college degree in letters and in the natural sciences. Just as Geoffroy had done in his time, Dardaud as a young adult went to Egypt for professional reasons. He would experience the tribulations of an explosive era. At twenty-five years of age, he became a permanent special envoy for the daily paper, *L'Intransigeant*. In 1936, at the age of thirty-seven, he became editor in chief of the French language paper, *Le Progrès égyptien*. During the Libyan desert campaign in 1941, he was a war correspondent with the 8th British Army. In 1944 he became director of *Agence France-Presse* for the Middle East in Cairo. In 1956 he was expelled from Egypt after the attack in Port-Said and the Suez Canal by Franco-British forces. He had a lot on his plate at the time. At that point, Dardaud was giving talks on Zarafa, which he subsequently published without wide distribution.

January 18, 1957: Dardaud's research snowballed. Professor Pierre Thévenard asked for permission to go and film Zarafa in the La Rochelle museum. In collaboration with the Paris museum, he was preparing to shoot a film on the story of the first giraffe in France: "I take inspiration for this work from the published talk by Mr. Dardaud, whose letter confirmed to me that you in fact possessed the remains of the celebrated giraffe of 1827."[12]

It had taken the combined efforts of a prefect and a scientist to bring Zarafa to the throne in France. One hundred thirty years later, it would take those of a journalist-writer and a doctor-filmmaker to revive the memory of her.

Pierre Thévenard had begun his career as a urologist (during Zarafa's time, he would have been able to help Geoffroy). In 1934 he assisted with the making of surgical films. In 1940 he decided, while awaiting the return of his clientele dispersed during the war, to use his talents through cinema to make scientific films. In 1941 he obtained the card for producer of special films, which allowed him to acquire film stock. He joined the association *Artisans d'art du cinéma*. Then he made a series of often award-winning productions, destined for research, for teaching, or for the public at large. In 1955 he published *La Radiocinématographie directe en 35 mm appliquée à la recherche* [Direct Radiocinematography for thirty-five mm cameras applied to research], a new cinematic technique that he perfected at the Institut Pasteur. While interested in Zarafa, he had just obtained the prize for a documentary at the Venice Film Festival. Officially presented on May 8, 1958, his film, *La Girafe à Paris*, was a short film (sixteen min.) shot in thirty-five mm color, both playful and didactic, utilizing the technique of animation.

After his expulsion from Egypt at the end of 1956, Gabriel Dardaud was appointed director of *Agence France-Presse* in Beirut, and then correspondent for *France-Soir* and *Europe 1*, where he then became one of the great voices covering the long Lebanese civil war. It was hardly surprising to note that these events interrupted the editorial projects of this eminent Zarafist. But in 1984 a great exhibit was organized on the topic of his heroine, with his collaboration, at the Île-de-France museum in the Château de Sceaux. The following year, Dardaud,

CHAPTER 9

after fifty-seven years spent in the Middle and Near East, left Beirut, after the looting of his apartment by a militia of Muslim fundamentalists, and went to live in Paris. He then published the result of his research begun thirty-six years earlier and which responded to the question posed to the curator in La Rochelle:

> Dr. E. Loppé, after a long inquiry, was able to affirm that the item given to the Museum in Verdun could only be the giraffe *said to be Daubenton's*, exhibited in the Museum in Paris in 1820 by the taxidermist Delalande. Professor Lacroix [from Verdun] had confused this South-African female, the first giraffe skin sent to the Museum in Paris, with the first live giraffe brought to France. The La Rochelle giraffe is well and truly the animal of King Charles X, gift from the Pasha of Egypt.[13]

Neither *Maasai,* nor *reticulated* in type, Zarafa, who hailed from East Sudan, is classified according to current zoological knowledge as a *Nubian giraffe*. Her Latin name is *giraffa cameloparladis cameloparladis*. Twice. In effect, the giraffe of all giraffes (figure 9.3).

And isn't it wonderful to consider that an animal so reputed for her mutism can still be talked about so much?

Figure 9.3. The Giraffe in all her majesty.
In d'Orbigny, *Dictionnaire universel d'histoire naturelle* © Muséum d'Histoire naturelle de La Rochelle / Lézard Graphique.

NOTES

FOREWORD

1. Muller, Z., et al., *Giraffa camelopardalis* (amended version of 2016 assessment), IUCN Red List of Threatened Species, 2018, https://www.iucnredlist.org/species/9194/136266699.

CHAPTER 1: THE GIRAFFE BEFORE ZARAFA: A POLYMORPHOUS MONSTER

1. Moses, Deuteronomy 14:3–5.
2. Agatharchides, *Sur la mer d'Erythrée* (Paris: Henri Étienne, 1557).
3. Oppian, *The Four Books of the Venerable Oppian, Greek Poet of Anazarba*, trans. into French by Florent Chrestien (Paris, 1579). Rendered into English by the translator of the present work.
4. Héliodore, *Ethiopiques, ou des amours de Théogène et Chariclée*, trans. Amyot, ed. M. Trogon. Rendered into English by the translator of the present work.
5. Pliny the Elder, *Natural History, Book 8: Land animals: their characteristics. The giraffe; when first seen at Rome*, trans. W. H. S. Jones (London: William Heinemann, 1949). Rendered into English by the translator of the present work.
6. Horace, *Epistolarum lib. II, epist. I*, trans. Monfalcon. Rendered into English by the translator of the present work.
7. Marco Polo, *Les Voyages . . . par Marc-Paul, . . .* (La Haye, 1735), vol. III, chapter XLI.
8. Léon l'Africain, *Description de l'Afrique* (Lyon, 1556).
9. Conrad Gesner, *Conradi Gesneri historia animalium. De quadrupedibus* (1620).

NOTES

10. Pierre Belon, *Les observations de plusieurs singularitez et choses mémorables trouvées en Grèce, Asie, Judée, Égypte, Arabie, et autres pays estranges* (Paris, 1588).

11. Marmol, *L'Afrique*, trans. Nicolas (Paris, 1667). Rendered into English by the translator of the present work.

12. Jan Jonston, *Historiae naturalis de quadrupedibus libri, cum aeneis figuris Johannes Jonstonus, medicinae doctor cocinnavit* (Amstelodami, Apud Joanem Jacobi Fil. Schipper, 1657).

13. Michel Baudier, *Histoire générale du sérail* [General history of the harem] (Lyon, 1659).

14. Pierre Richelet, *Dictionnaire* (1688).

15. Denis Didérot and Jean Le Rond d'Alembert, *Recueil de planches, sur les sciences, les arts libéraux et les arts méchaniques, avec leur explication* (Paris: Briasson, 1768).

16. Georges-Louis Buffon, *Histoire naturelle générale et particulière servant de suite à l'histoire des animaux quadripèdes par M. le comte de Buffon, intendant du Jardin et du Cabinet du Roi, de l'Académie française, de celle des sciences, etc.* (Paris: Impr. Royale, 1776), 3: 330.

17. Gabriel Dardaud, *Une girafe pour le roi* [A giraffe for the king] (Paris: Éditions Dumerchez-Naoum, 1985).

CHAPTER 2: ILLUSTRIOUS GODPARENTS: A PREDESTINED ANIMAL

1. Isidore Geoffroy Saint-Hilaire, *Vie, travaux et doctrine scientifique d'Étienne Geoffroy Saint-Hilaire* (Paris: Bertrand, 1847).

2. The National Convention was the first government of the French Revolution. Created after the great insurrection of August 10, 1792, it was the first French government organized as a republic, abandoning the monarchy altogether.

3. Georges Douin, *L'Égypte de 1802 à 1804: Correspondance des consuls de France en Égypte* (Cairo, 1925).

4. Sylvie Guichard, *Lettres de Bernardino Drovetti, consul de France à Alexandrie, 1803–1830* (Paris: Maisonneuve et Larose, 2003).

5. Cf. Dardaud, op. cit. Drovetti's correspondence has been conserved in the museum of Turin. It was in part analyzed and made known by the researcher Giovanni Marro in *Il corpo epistolare di Bernardino Drovetti ordinato ed illustrato* (Roma, 1940) and in "La fauna e la flora africana inviata in Europa da B. Drovetti dal 1805 al 1829" (*Bollettino dell'Istituto e Museo di Zoologia dell'Università di Torino*, 1949).

6. Gaston Wiet, *Les Consuls de France en Égypte sous le règne de Mohammed Ali*, éd. de la *Revue du Caire* (Le Caire, 1943).

7. Jean-François Champollion, *Lettres et journaux écrits pendant le voyage d'Égypte*, collected and annotated by Hermine Hartleben (Paris, 1986).

NOTES

CHAPTER 3: THE PASHA'S GIFT: DIPLOMACY AT STAKE

1. Victor Hugo, "L'Enfant," in *Les Orientales* (Paris, 1829). Rendered into English by the translator of the present work.
2. Gustave Loisel, *Histoire des ménageries de l'Antiquité à nos jours* [History of zoos from antiquity to today] (Paris, 1912), 3:137.
3. Cf. Dardaud, op. cit. The author, without providing the source of this information, wrote that the consul had mentioned the idea to the pasha. If the journalist found it only in Drovetti's correspondence, then the obsequious character of the latter leads us to be wary of his assertions on this subject.
4. *Notice sur la giraffe, envoyée au Roi de France par le pacha d'Égypte*, by Mr. Salze of the Académie des Sciences de Marseille, published in *Le Gazette universelle de Lyon*, June 14, 1827, republished in Paris in the *Moniteur universel*, June 30, 1827.
5. Either the giraffes were twins—which is extremely rare but possible, knowing that there was a major risk of premature death of one of them, as in June 2003, in England, at Maxwell Zoo—or they had the same mother, which implies that they were at least sixteen months apart in age, taking into account the time required for gestation. Knowing that Zarafa was six months old when she left Cairo, she could only have been the youngest and the other older by nearly two years. However, no account mentions the disparity in age between the two animals. One may then reasonably conclude that they were of two distinct mothers.
6. Michael Allin, *La Girafe de Charles X* (Paris: J.-C. Lattès, 2000). The author conducted research onsite to determine a plausible Egyptian itinerary.
7. *Notice sur la giraffe, envoyée au Roi de France par le pacha d'Égypte*, by Mr. Salze of the Académie des Sciences de Marseille, published in *Le Gazette universelle de Lyon*, June 14, 1827, republished in Paris in the *Moniteur universel*, June 30, 1827.
8. Cf. Dardaud, op. cit.
9. Cf. Dardaud, op. cit.

CHAPTER 4: ZARAFA IN FRANCE: A ROYAL TRIUMPH

1. Registry of the port in Marseille, declaration of Captain Manara, October 23, 1827.
2. Cf. Allin, op. cit. Unfortunately, Mr. Allin did not cite his sources. Neither did he mention Gabriel Dardaud as other than a "European journalist," while the latter had considerably defined the research on Zarafa.
3. Letter from Consul Drovetti to Pierre Barthalon, dated September 21, 1826, from Alexandria.
4. A. B. Clot-Bey, *De la peste observée en Égypte* (Paris, 1840).
5. Cf. Allin, op. cit.
6. Christophe (comte de) Villeneuve-Bargemon, *Statistique du département des Bouches-du-Rhône* (Marseille, 1826).

NOTES

7. Letter from Consul Drovetti to Pierre Barthalon, dated September 21, 1826, from Alexandria.
8. Letter from Consul Drovetti to Pierre Barthalon, dated September 28, 1826, from Alexandria.
9. Cited by Félix-L. Tavernier, "Deux hôtes princiers à l'Hôtel Roux de Corse sous la Restauration," *Revue illustrée du Musée du Vieux Marseille*, April 1959.

CHAPTER 5: THE ANIMAL UNDER A MAGNIFYING GLASS: AN ABERRATION OF NATURE

1. Étienne Geoffroy Saint-Hilaire, article, *Annales des Sciences naturelles* (Paris, 1827).
2. Article, *Gazette universelle de Lyon*, May 27, 1827.
3. In 1515 Albrecht Dürer made the famous wood engraving *Rhinocéros* from a sketch of the animal.

CHAPTER 6: BETWEEN SCIENCE AND RELIGION: AN ANIMAL AT A PRICE

1. Étienne Geoffroy Saint-Hilaire, *Sur la girafe* (1827).
2. Jean-Baptiste (de) Lamarck, *Philosophie zoologique* (Paris, 1809).
3. *Le Caducée: Souvenirs marseillais, provençaux et autres*, vol. II (Marseille, 1879).
4. Speech given at the *Académie* in Aix on December 11, 1934, by Mr. Édouard Aude, curator at the Méjanes library.
5. Unscrupulous merchants took advantage of the giraffe's passage to raise their prices.
6. *Gazette universelle de Lyon*, Wednesday, June 6, 1827.
7. Ibid., Friday, June 8, 1827.
8. Municipal archives of Lyon, letter to the chief of police, Wednesday, June 6, 1827.
9. *Gazette universelle de Lyon*, Saturday, June 9, 1827.
10. *Dernière Notice sur la Girafe, contenant la relation de son voyage à Saint-Cloud* (Paris: imp. Moreau, 1827).
11. Guy Barthélemy, *Les Jardiniers du Roy, petite histoire du Jardin des Plantes* (Paris: Editions Le Pélican, 1979).
12. Cf. Allin, op. cit.

CHAPTER 7: TRADE AND GIRAFFOMANIA: MARKETING MATERIALS

1. *Le Caducée*, op. cit.
2. *Gazette Universelle de Lyon*, July 5, 1827.

3. *Gazette Universelle de Lyon*, July 4, 1827.
4. *Journal des Dames et des Modes*, July and August 1827.
5. *Gazette universelle de Lyon*, August 3, 1827.

CHAPTER 8: POLITICAL CARICATURE: A SATIRICAL WEAPON

1. Cf. Dardaud, op. cit.
2. *Gazette Universelle de Lyon*, July 18, 1827.
3. L. S. Lambourne, "A Giraffe for George IV," *Country Life*, December 2, 1965.
4. Cf. Dardaud, op. cit. The author seems not to have taken this information into account as he wrote, "The giraffe, brought to England in August 1827, died several months later in Windsor."
5. Cf. Dardaud, op. cit.
6. Giuseppe Acerbi, Austrian Consul in Egypt from 1826 to 1834, tells the story of the Viennese giraffe in *Biblioteca Italiana* vol. 50 (1828).

CHAPTER 9: ZARAFA'S LEGACY: AN EXISTENTIAL FABLE

1. Cited by Claude Drouin (unrevealed sources), in his thesis, *La Girafe de Charles X, une révélation pour les Français* [The Giraffe of Charles X, a Revelation for the French], defended November 26, 1987, at the École nationale vétérinaire in Lyon.
2. Richard Owen, "On the Birth of the Giraffe at the Zoological Society's Gardens," *Transactions of the Zoological Society*, vol. III, p. 175.
3. Paul Gervaix, *Histoire naturelle des mammifères* (Paris, 1854).
4. François Charles-Roux, "Clot-Bey et le consul Cochelet," *Cahiers d'Histoire égyptienne*, série II, no. 2 et 3, 1950.
5. N. Joly and A. Lavocat, *Recherches historiques, zoologiques, anatomiques et paléontologiques sur la girafe* (Strasbourg, 1845).
6. Cf. Dardaud, op. cit. The author cites R. Didier and A. Boudarel, *L'Art de la taxidermie au XXème siècle* (Paris: Lechevalier, 1948).
7. *Le Caducée, souvenirs marseillais, provençaux et autres*. Tome II (Marseille, 1879).
8. Cf. Dardaud, op. cit.
9. Letter, n.d., typeset then corrected by hand; archives, *Muséum d'Histoire naturelle de La Rochelle*.
10. Napoleon's dromedary was later transferred to the Napoleonic museum on the island of Elba and then moved to that of Île-d'Aix, where it remains.
11. Archives, *Muséum d'Histoire naturelle de la Rochelle*.
12. Cf. Archives, *Muséum d'Histoire naturelle de la Rochelle*.
13. Cf. Dardaud, op. cit.

BIBLIOGRAPHY

IN ALPHABETICAL ORDER BY AUTHOR

Acerbi, Giuseppe. *Biblioteca Italiana*. 1828.
Agatarchides. *Sur la mer d'Erythrée*. Paris: Henri Étienne, 1557.
Albert-le-Grand. *De animalibus*. Edited by Father Jammy. Lyon, 1651.
Aldrovandi, Ulysse. *Apud Jo Baptista Ferronii*. Bonon, 1642.
Allin, Michael. *La Girafe de Charles X*. Translated by Stéphane Carn. Paris: J. C. Lattès, 2000.
Barthélemy, Guy. *Les Jardiniers du Roy, petite histoire du Jardin des Plantes*. Paris: Editions Le Pélican, 1979.
Baudier, Michel. *Histoire générale du serrail*. Lyon, 1659.
Belon, Pierre. *Les Observations de plusieurs singularitez et choses mémorables trouvées en Grèce, Asie, Judée, Égypte, Arabie, et autres pays estranges*. 3 volumes. Paris, 1588.
Bochart. *Hierozoïcon, sive bipartitum opus de Animalibus scripturae*. Lugduni Batavorum, 1712.
Breydenbach, Bernard (von). *Opusculum sanctarum peregrinto rium*. 1486.
Buffon, Georges-Louis. *Histoire naturelle générale et particulière servant de suite à l'histoire des animaux quadrupèdes par M. le Comte de Buffon, intendant du Jardin et du Cabinet du Roi, de l'Académie française, de celle des sciences, etc.* Paris: Imprimerie Royale, 1776–1787.
Champollion, J. F. *Lettres et journaux écrits pendant le voyage d'Égypte*. Collected and annotated by H. Hartleben. Paris, 1986.
Clot-Bey, A. B. *De la peste observée en Égypte*. Paris, 1840.
Cuvier, Frédéric. *Dictionnaire des Sciences Naturelles*. Planches, 2e partie. Paris: F. G. Levrault, 1818–1829.

BIBLIOGRAPHY

Cuvier, Frédéric, and Étienne Geoffroy Saint-Hilaire. *Histoire naturelle des mammifères*. Paris, 1842.
Cuvier, George. *Règne animal*. Paris, 1817.
Cuvier, George. *Histoire des progrès des sciences naturelles*. Paris, 1836.
Cuvier, George. *Leçons d'anatomie comparée*. Paris, 1837.
Dardaud, Gabriel. *Une girafe pour le roi*. Paris: Editions Dumerchez-Naoum, 1985.
Darwin, Charles. *L'Origine des espèces*. Paris: Reinwald & Cie. 1859.
Diderot, Denis, and Jean Le Rond d'Alembert. *Recueil de planches, sur les sciences, les arts libéraux et les arts méchaniques, avec leur explication*. Paris: Briasson, 1768.
Didier, R., and A. Boudarel. *L'Art de la taxidermie au XXème siècle*. Paris: Lechevalier, 1948.
Douin, G. *L'Égypte de 1802 à 1804. Correspondance des consuls de France en Égypte*. Cairo, 1925.
Fourier, Charles. *Oeuvres complètes*. Paris: La Phalange, 1841–1845.
Geoffroy Saint-Hilaire, Étienne. *Sur la girafe*. 1827.
Geoffroy Saint-Hilaire, Isidore. *Dictionnaire classique d'histoire naturelle*. Paris, 1823.
Geoffroy Saint-Hilaire, Isidore. *Vie, travaux et doctrine scientifique d'Étienne Geoffroy Saint-Hilaire*. Paris: Editions Bertrand, 1847.
Gervaix, Paul. *Histoire naturelle des mammifères*. Paris, 1854.
Gesner, Conrad. *Conradi Gesneri historia animalium. De quadrupedibus*. 1620.
Guichard, Sylvie. *Lettres de Bernardino Drovetti, consul de France to Alexandrie 1803–1830*. Paris: Maisonneuve and Larose, 2003.
Haug, Hans. *L'Art populaire en France—La Girafe de Charles X: son influence sur l'art populaire et sur la mode*. Paris: Istra, 1932.
Héliodore. *Éthiopiques, ou des amours de Théagène et de Chlariclée*. Translated into French by Amyot. Edited by M. Trogon, 1584. Rendered into English by the translator of the present work.
Horace. *Epistolarum lib. II, epist. I*. Translated into French by Monfalcon. Paris and Lyon, 1834. Rendered into English by the translator of the present work.
Hugo, Victor. *Les Orientales*. Paris, 1829.
Joly, N. *Notice sur l'histoire, les moeurs et l'organisation de la girafe*. Toulouse, 1844.
Joly, N., and A. Lavocat. *Recherches historiques, zoologiques, anatomiques et paléontologiques sur la girafe*. Strasbourg, 1845.
Jonston, Jan. *Historiae naturalis de quadrupedibus libri, cum aeneis figuris Johannes Jonstonus, medicinae doctor concinnavit*. Amstelodami: Apud Joannem Jacobi Fil. Schipper, 1747.
Laissus, Y., and J.J. Petter. *Les Animaux du Muséum*. Paris: Imprimerie Nationale, 1993.
Lamarck, J. B. (de). *Philosophie zoologique*. Paris: Dentu, 1809.
Léon l'Africain. *Description de l'Afrique*. Lyon, 1556.
Levaillant, François. *Premier voyage dans l'intérieur de l'Afrique par le cap de Bonne-Espérance*. Paris, 1790.

Levaillant, François. *Second voyage dans l'intérieur de l'Afrique.* Paris, 1796.
Linné, C. *Caroli Linnaei Systema naturae.* Lugduni, 1789.
Loisel, Gustave. *Histoire des ménageries de l'Antiquité à nos jours.* (Paris, 1912), 3:137.
Marmol. *L'Afrique.* Translated by Nicolas. Paris, 1667. Rendered into English by the translator of the present work.
Marro, Giovanni. *Il corpo epistolare di Bernardino Drovetti ordinato et illustrato.* Roma, 1940.
Marro, Giovanni. "La fauna e la flora africana inviata in Europa da B. Drovetti dal 1805 al 1829." *Bolletino dell'Istituto e Museo di Zoologia dell'Universita di Torino,* 1949.
Moses. Deuteronomy 14:3–5. *Bible.*
Nimier, Marie. *La Girafe.* Paris: Editions Gallimard, 1989.
Oppian. *The Four Books of the Venerable Oppian, Greek Poet of Anazarba.* Translated into French by Florent Chrestien. Rendered into English by the translator of the present work. Paris, 1579.
Paré, Ambroise. *Les oeuvres d'Ambroise Paré, . . . 8e édition, revues et corrigées en plusieurs endroicts et augmentées d'un fort ample traicté des fiebvres . . . nouvellement trouvé dans les manuscrits de l'autheur.* Paris: G. Buon, 1585.
Pline. *Histoire naturelle, Livre VIII: Animaux terrestres: leurs caractères. De la girafe: quand la première parut à Rome.* Translated by Émile Littré. Paris: Panckoucke, 1829.
Polo, Marco. *Les Voyages très-curieux et très-remarquables achevés par toute l'Asie, Tartarie, Mangi, Japon, les îles orientales, îles adjacentes et l'Afrique, commencés en l'an 1252 par Marc-Paul, Vénitien, historien remarquable par sa fidélité, le tout divisé en trois livres.* La Haye, 1735.
Richelet, Pierre. *Dictionnaire.* 1688.
Stendhal. *Correspondance 1821–1834.* Paris: La Pléïade.
Thevet, André. *Cosmographie du Levant.* Lyon, 1554.
Villeneuve-Bargemon, Christophe (comte de). *Statistique du Département des Bouches-du-Rhône.* Marseille, 1826.
Wiet, Gaston. *Les Consuls de France en Égypte sous le règne de Mohammed Ali.* Cairo: Ed. de la *Revue du* Caire, 1943.
Wynants, M. *La girafe du Roi Charles X.* Album du petit berger. Bruges: Brouwer, 1961.

CORRESPONDENCE (BY LETTER IN CHRONOLOGICAL ORDER)

Note: These letters relate to the following individuals, noted in parentheses as follows:

Pierre Barthalon (Barthalon)
Bernardino Drovetti, French Consul in Egypt (Drovetti)
Baron de Damas, Minister of Foreign Affairs (Damas)
Bottu, Agent of Foreign Affairs in Marseille (Bottu)

BIBLIOGRAPHY

Christophe de Villeneuve-Bargemon, Prefect of the Bouches-du-Rhône (Villeneuve)
Public Health Officials in Marseille (Intendants)
M. Bosc, Veterinarian at the Museum of Natural History (Bosc)
Count of Corbières, Minister of the Interior (Corbières)
Professor-Administrators of the Museum of Natural History in Paris: G. Cuvier, L. Cordier, and Bosc (Professors)
Étienne Geoffroy Saint-Hilaire, Museum professor (Geoffroy)

The geographical source of the letter is indicated in parentheses where possible.
Almost all of the collections were found in the Archives départementales des Bouches-du-Rhône.
Others were found within private collections.
The rest, noted with an asterisk, are found in the Library of the Museum of Natural History in Paris.
If a letter contains no date, the letter is summarized to justify its placement in the text's chronology.

Letter from Damas to Corbières: The Minister of Foreign Affairs announces the departure of the giraffe from Egypt.*
21 September 1826, Letter, Drovetti to Barthalon (Alexandria).
28 September 1826, Letter, Drovetti to Barthalon (Alexandria).
29 September 1826, Letter, Drovetti to Bottu (Alexandria).
10 October 1826, Letter, Damas to Bottu (Paris).
26 October 1826, Letter, Bottu to Villeneuve (Marseille).
27 October 1826, Letter, the Intendants to Villeneuve (Marseille).
28 October 1826, Letter, Villeneuve to the Intendants (Marseille).
30 October 1826, Letter, Villeneuve to the Intendants (Marseille).
8 November 1826, Letter, Villeneuve to Corbières (Marseille).
Letter, Villeneuve to the Customs Director (Marseille).
16 November 1826, Letter, Consul of Sardinia to Villeneuve (Marseille).
Instructions from Bosc (Paris)
18 November 1826, Letter, Villeneuve to Corbières (Marseille).
21 November 1826, Muséum administration: excerpt from a report to Corbières (Paris).
23 November 1826, Letter, Villeneuve to Corbières (Marseille).
24 November 1826, Letter, Villeneuve to the Customs Director and to the Mayor of Marseille (Marseille).
28 November 1826, Letter from the Professors to Villeneuve (Paris).
28 November 1826, Letter from the *Collège* Principal to Villeneuve (Marseille).
5 January 1827, Letter, Villeneuve to the Professors (Marseille).
19 January 1827, Letter from the Professors to Villeneuve (Paris).

BIBLIOGRAPHY

6 February 1827, Letter, Villeneuve to the Professors (Marseille).
12 February 1827, Letter, Villeneuve to the Professors (Marseille).
3 March 1827, Letter, Villeneuve to the Commander of the Military Police (Marseille).
9 March 1827, Letter, Drovetti to Villeneuve (Alexandria).
15 March 1827, Letter from the Professors to Villeneuve (Paris).
19 March 1827, Letter, Villeneuve to the Professors (Marseille).
26 March 1827, Letter, Villeneuve to the Professors (Marseille).
5 April 1827, Letter from the Professors to Villeneuve (Paris).
27 April 1827, Letter, Villeneuve to Count Félix d'Albertas (Marseille).
28 April 1827, Letter, Villeneuve to the Professors (Marseille).
Invoice: For the giraffe's raincoat (Marseille).
4 May 1827, Letter, Villeneuve to Count Félix d'Albertas (Marseille).
17 May 1827, Circular from Villeneuve to the mayors of the department (Marseille).
Letter, Villeneuve to the Commander of the Military Police.
17 May 1827, Letter, Villeneuve to the Professors (Marseille).
18 May 1827, Letter, Villeneuve to the subprefect, Aix-en-Provence, to the prefects of the departments of Vaucluse, Drôme, Isère, etc. (Marseille).
18 May 1827, Proposal from J. B. Chapsal to Villeneuve (Marseille).
19 May 1827, Letter, Geoffroy to Villeneuve (Marseille).
21 May 1827, Letter, Geoffroy to Villeneuve (Aix-en-Provence).
23 May 1827, Letter from the subprefect of Aix-en-Provence to Villeneuve (Aix-en-Provence).
23 May 1827, Letter, Villeneuve to the prefects of Vaucluse, Drôme, and Isère (Marseille).
24 May 1827, Letter from the Commander of the Military Police to Villeneuve (Lambesc).
24 May 1827, Letter, Geoffroy to Villeneuve (Désiré, place between Saint-Andiol and the Durance).
28 May 1827, Letter, Villeneuve to Corbières, with a copy to the Professors (Marseille).
2 June 1827, Letter, Geoffroy to Villeneuve (Lyon).
5 June 1827, Letter from the Professors to Villeneuve.
5 June 1827, Letter, Geoffroy to Corbières.
8 June 1827, Letter, Geoffroy to Villeneuve (Lyon).
12 June 1827, Letter, Corbières to Villeneuve (Paris).
20 June 1827, Letter, Villeneuve to the Professors (Marseille).
23 June 1827, Letter from Isidore Geoffroy St-Hilaire to his mother (Auxerre).
27 June 1827, Letter, Geoffroy to Cuvier.*
Letter from the Professors to Corbières: they provide the itinerary of the giraffe's procession between Paris and St-Cloud to the Minister of the Interior.
4 July 1827, Letter from the Professors to Villeneuve (Paris).
12 July 1827, Letter, Geoffroy to Villeneuve.
4 September 1827, Letter, Villeneuve to Geoffroy (Marseille).
22 October 1827, Letter, Geoffroy to Villeneuve (Paris).

BIBLIOGRAPHY

PERIOD DOCUMENTS DEVOTED TO ZARAFA (IN ALPHABETICAL ORDER BY TITLE)

À la giraffe, invocation avec choeurs et couplets, paroles et musique de F.G. avec lithographie par Langlumé. Paris, 1827.

Almanach stéréotype à la girafe pour 1829, Véritable messager boiteux to la girafe. New editions in 1830, 1831, 1845. . . .

Cantate avec choeurs, set to the melody of "Cadet-Roussel".

Dame Girafe à Paris, ou Aventures et voyage de cette illustre étrangère, racontés par elle-même, en réponse au discours de S. E. l'Ours Martin; avec le détail des fêtes que lui ont données les pensionnaires du Jardin du Roi, à-propos historique, précédé d'une dissertation scientifique par Bufon [sic] à l'usage des visiteurs de la Ménagerie. By Charles-François Berlu, 1827.

Dernière notice sur la Girafe, contenant la relation de son voyage to Saint-Cloud. Paris: Moreau, 1827.

Discours de la Girafe au chef des Six Osages (ou Indiens), prononcé le jour de leur visite au Jardin du Roi. Translated from Arabic by Alibassan, the giraffe's interpreter. Paris: Balzac, 1827.

Épitre à la Girafe. By D. Charles.

La girafe. Printed sheet.

La girafe à Saint-Cloud, romance, sur une musique de M. Ed. Hardy. Sold for 1.5 francs in Paris, at the music shop of A. Petibon, rue du Bac, N° 31. By the Countess of Oglou.

La Girafe ou le gouvernement des bêtes, divertissement interrompu donné par MM. Les Animaux du Jardin du Roi comme un témoignage de leur reconnaissance envers le Pacha d'Égypte, à l'occasion de l'arrivée de la Girafe à la Ménagerie de Paris. Paris, 1827.

La girafe, ou une journée au Jardin du Roi, tableau à propos en vaudeville de Théaulon de Lambert et Th. Anne et Gondolier, représenté pour la première fois à Paris le 7 juillet 1827 sur le théâtre du Vaudeville. Paris, 1827.

Le Petit Chansonnier des Dames et des Demoiselles. Castiaux, 1828.

Les adieux de la girafe, valse de Singer pour piano, divertissement to la guitare. By F. Carulli.

Lettre, la Girafe au Pacha d'Égypte, pour lui rendre compte de son voyage to Saint-Cloud, et renvoyer les rognures de la censure de France au journal qui s'établit to Alexandrie en Afrique. By Salvandy (comte de), Narcisse-Achille. Paris, 12 July 1827.

Nouvelle notice sur la girafe, avec en frontispice, une lithographie de Renou. Ferlus, 1827.

Observations faites sur la girafe envoyée au Roi par la Pacha d'Egypte, et sortie du lazaret de Marseille le 14 novembre 1826. Memoires, Muséum National, Salze, 1826.

Seconde Lettre, la Girafe au Pacha d'Égypte, en lui envoyant son album enrichi des dernières noirceurs de la censure, Salvandy (comte de), Narcisse-Achille. Paris, 8 August 1827.

Tablettes de la girafe du Jardin des Plantes. Lettre à son amant du désert, extrait de La Vie privée et publique des animaux, vignettes par Grandville, Scènes de la vie privée et publique des animaux. By Charles Nodier. 1867.

BIBLIOGRAPHY

Trois fables sur la giraffe. By Adolphe Jauffert. Paris and Marseille, 1827.

Un autre monde. Transformations, visions, incarnations, ascensions, locomotions, explorations, pérégrinations, excursions, stations, cosmogonies, fantasmagories, rêveries, folâtreries, facéties, lubies, métamorphoses, zoomorphoses, lithomorphoses, métempsycoses, apothéoses et autres choses, Text by Taxile Delord, Paris: H. Fournier, 1844.

NEWS ARTICLES (IN CHRONOLOGICAL YEAR OF APPEARANCE)

1826, 21 November. *Le Messager*, Méry.
1827, 14 February. *La Gazette universelle de Lyon.*
1827, 21 February. *La Gazette universelle de Lyon.*
1827, 27 May. *La Gazette universelle de Lyon.*
1827, 6 June. *La Gazette universelle de Lyon.*
1827, 8 June. *La Gazette universelle de Lyon.*
1827, 9 June. *La Gazette universelle de Lyon.*
1827, 10 June. *La Gazette universelle de Lyon.*
1827, 10 June. *Le Moniteur universel.*
1827, 13 June. *Le Figaro.*
1827, 13 June. *Le Moniteur universel.*
1827, 14 June. *Le Figaro.*
1827, 14 June. *La Gazette universelle de Lyon.*
1827, 16 June. *La Gazette universelle de Lyon.*
1827, 17 June. *Le Figaro.*
1827, 26 June. *La Gazette universelle de Lyon.*
1827, 27 June. *La Gazette universelle de Lyon.*
1827, 30 June. *Le Moniteur universel.*
1827, 2 July. *La Gazette universelle de Lyon.*
1827, 2 July. *Le Moniteur universel.*
1827, 3 July. *Le Figaro.*
1827, 4 July. *La Gazette universelle de Lyon.*
1827, 4 July. *Le Moniteur universel.*
1827, 5 July. *La Gazette universelle de Lyon.*
1827, 5 July. *Le Globe.*
1827, 6 July. *Le Figaro.*
1827, 6 July. *La Gazette universelle de Lyon.*
1827, 6 July. *La Gazette de France.*
1827, 6 July. *Le Précurseur.*
1827, 8 July. *Le Figaro.*
1827, 8 July. *La Gazette universelle de Lyon.*
1827, 9 July. *Le Figaro.*

BIBLIOGRAPHY

1827, 10 July. *Le Figaro*.
1827, 10 July. *Le Moniteur universel*.
1827, 11 July. *La Gazette universelle de Lyon*.
1827, 11 July. *Le Moniteur universel*.
1827, 15 July. *La Gazette universelle de Lyon*.
1827, 18 July. *La Gazette universelle de Lyon*.
1827, 20 July. *Le Figaro*.
1827, 27 July. *Le Figaro*.
1827, 28 July. *Le Figaro*.
1827, 29 July. *La Gazette universelle de Lyon*.
1827, 2 August. *Le Figaro*.
1827, 3 August. *La Gazette universelle de Lyon*.
1827, 4 August. *Le Figaro*.
1827, 7 August. *Le Figaro*.
1827, July and August. *Journal des Dames et des Modes*.
1827, 14 August. *La Gazette universelle de Lyon*.
1827, 21 August. *La Gazette universelle de Lyon*.
1827. Mongez. "Mémoire sur les animaux promenés ou tués dans les cirques." *Annales des sciences naturelles*. Paris.
1830, 17 June. *Étude de la philosophie, morale sur les habitudes du jardin des plantes*, illustrated newspaper *Silhouette*.
1833, 28 September. *Le Garde national*, Méry.
1879. *Le Caducée: Souvenirs marseillais, provençaux et autres*. Volume II. Marseille.
1950. "Clot-Bey et le consul Cochelet." *Cahiers d'Histoire Égyptienne*. Série II, N° 2 and 3. By François Charles-Roux.
1951, January. Dardaud, Gabriel. "L'Extraordinaire Aventure de la girafe du Pacha d'Égypte." *Revue des conférences françaises en Orient* 15, no. 1.
1957, October. Thévenard, Pierre. "La Girafe à Paris." *Naturalia*.
1959, April. Tavernier, F.-L. "Deux hôtes princiers à l'Hôtel Roux de Corse sous la Restauration." *Revue illustrée du Musée du Vieux Marseille*.
1959, May. Berthier de Sauvigny, Guillaume. "L'année de la girafe." *Historia*.
1963. Dagg, Anne. "A French giraffe." *Frontiers* (Academy of Natural Sciences, Philadelphia).
1965, 2 December. Lambourne, L. S. "A Giraffe for George IV." *Country Life*.
1970. Bolzinger, J. *La girafe de Charles X, L'Écho de Joigny*.
1973, 12 July. Cunningham Scriven, Sheila. "The Year of the Giraffe." *Country Life*.
1975, 16 April. *Le Figaro*.
1977, Autumn. Delohen, Pierre. "La girafe qui traversera Auxerre." *L'Yonne républicain*.
1984, 14 May. *Le Point*.
1984, April. Massard, D. "La girafe de Charles X" (comic). *Amis-Coop*, no. 268. Paris: SCATOCCE.
1984, October. Ellenberger, Michel. "La girafe du roi." *Pour la Science*.

BIBLIOGRAPHY

1995. Bonvarlet, X. "Une girafe au destin royal." *Civic*, no. 54.
1998, 28 July. Rozensweig, Luc. "La girafe, apôtre de la non-violence." *Le Monde*.
2000, 24 February. P. G. "Mais que faisait donc cette girafe to Paris?" *L'Événement du Jeudi*.
2003, 15 March. Fontaine, Jean-Pierre. "Le passage de la girafe." *Yonne Mag*.
2005, February. Hartengerger, Jean-Louis. "La girafe arrive!" *Les Génies de la Science*, no. 22.

ADDITIONAL DOCUMENTS (IN CHRONOLOGICAL ORDER)

Registre du port de Marseille, déclaration du capitaine Manara. 23 October 1827.
Rapport d'autopsie de l'antilope femelle, par le vétérinaire marseillais docteur Vial. 6 March 1827.
Rapports de police de Lyon. June 1827, archive municipale.
Catalogue d'août 1840 à décembre 1845: autopsie de Zarafa et répartition des divers éléments de son corps dans les différents laboratoires du Muséum National d'Histoire Naturelle.
Communication faite à l'Académie d'Aix, par le conservateur de la Bibliothèque Méjanes, Aude, Édouard. 11 December 1934.
Catalogue de vente à l'hôtel Drouot les 14 et 15 avril 1975, commissaire-priseur Maître Paul Renaud.
Catalogue de l'exposition *Une girafe pour le Roi*, du 19 avril au 15 juillet 1984 au Musée de l'Ile-de-France du Château de Sceaux.
Thesis by Drouin, Claude. *La Girafe de Charles X, une révélation pour les Français*. Defended November 26, 1987, at the École nationale vétérinaire de Lyon.

INDEX

Page references for figures are italicized.

Académie française, 107
Académie royale de Musique, 27
Académie des Sciences (Marseille), 55, 159n4, 159n7
Académie des sciences (Paris), 147
Acerbi, Guiseppe, 161n6
Acropolis, 81
Adana region, 134
Adrianople, 137; Treaty of, 138
Adriatic Sea, 137
Africa, 4, 59, 68, 104, 119, 123; Central, 15, 58; city (*see* Sennar); *Description of*, 4; Eastern, ix; elephant, 22, 151; exoticism, 55; expedition, 143; fauna, xii; giraffe(s), 9, 48, 62, 78, 88, 104, 154; landscape, 51, 89; North, 39, 152; people of, 7, 103, 149; Southern, ix, 10, 48, 55; Sub-Saharan, 152
Agasse, Jacques-Laurent, 126
Agatharchides, 2
Agence France-Presse, 152
Aix (-en-Provence), 71, 80–81, 85, 160n4; *Cérémonial of*, 76; mayor (*see* Estienne du Bourquet (d'), Louis-Jules); subprefect, 80
Akif, Reshid, 81
d'Albertas, Félix (count, marquis), 69–70, 75
Albion. *See* England
Alexander I, 113
Alexandria, 20–21, 34, 37–39, 45, 63–64, 123, 137; Geoffroy St. Hilaire's adventure, 18; library, 17
Algeria, 113, 148; affair, 137
Algiers, 113; dey of (*see* Hussein); occupation of, 143; port of, 137–138; regency of, 113, 143
al-Hasan ibn Muhammad al-Wazzan al-Fasi. *See* Leo Africanus
Ali, Ibrahim, 22, 27
Ali, Muhammad, 13–15, *14*, 21–23, 25, 27–28, 31, 33–35, 38, 40, 42, 48, 53, 64, 73, 76, 81, 109, 111–112, 115, 118, 123–126, 134, 145–146, 150, 152, 154, 159n3; son (*see* Ibrahim Pasha)

INDEX

Alibassan (pseudonym). *See* Balzac, Honoré de
Allin, Michael, xi, 159n6, 159n2
Alps, 137
al-Qazwini, Zakariya, 3
Amarindian poem, 129
Amathonte, 29
America, xii, 151; geneticists, 149; South, 41; tribe, 129; writer, xi
anabula, 4
Angoulême (Duchess of), 90, 116, 118, 121–2, 144
Angoulême (Duke of), 90, 118
Anne, 168
Anne. *See* Beaujeu, Anne de
Arabia, 15
Arc de triomphe, 44
Aristotle, 2
Arléa, xii
Arles' theater, 44
Arnay-le-Duc, 76
Arta, Gulf of, 144
Artois, Count of. *See* Charles X
Asia: East, 42; rhinoceros, 63; Southeast, 10
Athens, 123; fall of, 81, 92
Atir, 49, 51, 62, 75, 80, 112, 123, 125, 132, 136, 143, 145
Aubenton, Louis Jean-Marie d'. *See* Daubenton
Auberive, 85
Aude, Édouard, 160n4
Austerlitz (pont d'), 136
Austria, 90; consul, 161n6; giraffe, 137
Auxerre, 91–92
Avignon, 71, 80, 83, 85

Baar el Abial (White River). *See* Nile
Balkans, 25
Balocchi, Luigi, 27
Balthalon, Pierre, 31, 33

Balzac, Honoré de, 130, 143; Alibassan (pseudonym), 130
Barbaria. *See* (North) Africa
Bartsch, Jacob, 7
Basha. *See* Ali, Muhammad
Baudier, Michel, 7
Bazin, 78
Beastophile, Madame (fictional name), 115
de Beauharnais, Joséphine, 152
Beaujeu, Anne de, 10
Bedlam, 114
Bedouin. *See* el Berberi, Hassan.
Beirut, 153–154
Belgium, 61
Belleville, 90
Belon, Pierre, 4
Bengal, 4
Berlinghieri, Deniello, 93
Berlioz, Hector 28
Berry (Duchesse de), 43, 118, 122, 136
Bertholin (fictional names), 114
Bertu, Charles-François, 115
bey: Clot (*see* Clot-bey); Mamluk, 13; of Sudan (*see* Mouker Bey)
Beyle, Henri. *See* Stendhal
Bezançon, Rémi, xii
Biba (Sultan), 10
Big Dipper, 7
Blainville. *See* Ducrotay de Blainville, Henri-Marie
Bois de Vincennes, 152
Bonaparte, Caroline, 21
Bonaparte, Napoléon, 13, 16, 20, 26, 84, 135, 147, 152; armies, 26, 42; dromedary, 152, 161n10; Egyptian campaign, 67, 69, 73, 113, 149
Bonnet, François (Lord), 39
Bordeaux; college, 147; Duc of, 117
Bory de Saint-Vincent, Jean-Baptiste, 99
Bosc d'Antic, Louis-Augustin, 45, 49

INDEX

Botanical Garden(s). *See* King's Garden
Bottu (Agent of Foreign Affairs), 37, 39–40, 49
Bouches-du-Rhônes; department, 152; prefect of (*see* Villeneuve Bargemon, Christophe de); policemen, 81; *Statistics on the Department of* (*see* Villeneuve Bargemon, Christophe de)
Bourbon, Marie-Thérèse Charlotte de. *See* Angoulême (Duchess of)
Bourdelle, Édouard (professor), 151
Bourmont, Louis de (minister of war), 143
Boyer, Pierre François Xavier (general-baron), 38, 125
Brascassat, Jacques Raymond, 76
British; Army (8th), 153; beheaded citizens, 31; expedition of 1807, 21; fleet, 125
Buffon, Georges-Louis Leclerc de (comte), 7, *11*, 58, 68, 115, 147
Burgundy, 87, 90
Burkina Faso, ix
Byron, George Gordon (Lord), 25–27, 112; *The Oath of Lord Byron in Missolonghi*, 25–26; *Lord Byron on His Death-Bed*, 26
Byzantium, 28

Caesar, Julius (emperor), 3
Cairo, xi, 15, 31, 34, 53, 106, 118, 150, 152–153, 159n5; Insitute of Liberal Arts and Sciences, 17
Canning, George (minister), 113
Canut (revolts), 144
Cape Town, 97, 147
Capo d'Istria, Jean, 144
Caroline (queen). *See* Bonaparte, Caroline
Carulli, Ferdinando, 102
Cassiopeia, 7
Châlons (-sur-Saône), 84, 86–87, 91–92, 121

Chamberlain (Lord), 126, 129
Champollion, Jean-François, 22, 33, 137
Chapsal, Jean, 72
Chariclea. *See* Heliodorus of Emesa
Charles Felix of Sardinia, 22
Charles X, 23, 25, 31–3, *30*, 37, 44, 53, 64, 76, 85, 90, 109, 112–3, 116–8, 121–2, 124–5, 134, 136–8, 142–5, 148, 151–2, 154; government, 69, 81 the Charter, 33
Chateaubriand, François-René de, 32, 109, 129
Chenoise castle farm, 92
Chevaliers de la Foi. *See* Villeneuve Bargemon, Christophe de
Chicago, x
China: ancient, 4; people, 133; sovereign, 125
Chio / Chios, 25, *26*, 29, 115
Choisy, 92
Chouquet, Barthélémy, 70–71, 76, 80, 123
Clairefontaine earthenware, 101
Cléomène, 27
Clot, Antoine Barthélémy, 42, 146
Clot-Bey. *See above*
Collège de Navarre (Paris), 16
Collège royal de Marseille. *See* Marseille
Collège des Sciences (Paris), 147
Constantin I, 103
Constantinople, 7, 13, 21, 25, 37–38, 49, 92, 97, 122, 125, 137, 146
The Convention, 16, 158n2
Conyngham, Elizabeth (Lady), 126–129, *127, 128, 129*
Cook, James (Captain), 70
Corbière, Jacques-Joseph (Count, Minister of the Interior), 44, 47–48, 64, 75, 81, 83, 85–87, 90, 111, 118, 123

INDEX

Cordier, Pierre Louis Antoine (professor-administrator of the Museum), 49
Corinth, Gulf of, 27; *Siege of* (*see* Rossini)
Cortier, Maurice (captain), 1
Côte d'Or, 91
Crisp (doctor), 129
Cuvier, Frédéric, *56*, 147
Cuvier, Georges (professor), 16, 20, 22, 49, 66, 69, 92, 97, 118, 149, 151; step-daughter (*see* Duvaucel, Sophie); younger brother (*see next*)
Cynegetica. See Oppian

Damas, Ange Hyacinthe Maxence de (baron), 39–40
Danube Valley, 137
Dardaud, Gabriel, xi, 152–3, 158n5, 159n3, 159n2, 161n5
Darf(o)ur, 51, 112
Darwin, Charles, 149–50
Daubenton, 16; giraffe, 152, 154
Daumier, Honoré, 144
Delacroix, Eugène, 26–27
Delalande, Pierre Antoine, 152, 154
Delessert, Adolphe, 151g
Delft porcelain, 101
Democritus, 3
Denon, Dominique -Vivant, 21
Désiré, 79
Deuteronomy, 1
dey. *See* Hussein
Deval, Pierre (Consul of France in Algiers), 113
Diard, Pierre-Médard, 151
Diogenes, *100*
Dioum, Baba, x
Dipper. *See* Big Dipper
Doumergue, Gaston, 152
Draguignan, 43
Drôme, 79, 81, 85
Drovetti, Bernardino (Consul), 13, 20–23, *21*, 31–33, 35, 37–38, 40–41, 44, 48–49, 52, 64, 92, 123, 125, 134, 150, 159n3; nephew, 38, 44, 52, 73
Dublin zoo, 146
Ducrotay de Blainville, Henri-Marie, 150
Dumas, Alexandre, 51
Durance River, 79–80
Durant, 33
Duvaucel, Alfred, 151
Duvaucel, Sophie, 97

Ebed, Joseph / Youssef, 73, 75, 80, 123; father of, 73
the *Echo* (ship), 38
École des Beaux-Arts, 28
Effendis (school of), 119
Egypt, 13–17, 20–23, 33–34, 38–39, 47, 51, 70, 85, 124–126, 137, 145–8, 152–3; army, 26, 137; campaign of, 38, 42, 67, 73, 113, 152; collection, 33; cows, 34, 52, 70, 112, 116; *Description de l'Égypte*, 21, 22, *30;* Egyptology, 22, 137; Egyptophilia, 22; fauna, 16; fleet, 125, 134; giraffe, xii, 25, 33, 79, 116, 126, 129; government, 20, 31, 143, 146; health services, 42; involvement in Greece, 25, 109; keepers (of the giraffe), xi, 47–50, 71, 73, 77, 91, 112, 122, 124, 126; Legion, 73; liberation, 125; pasha of (*see* Ali, Muhammad); people of, 2, 15, 115, 118; refugees, 73; Sultan of, 10; Viceroy of (*see* Ali, Muhammad); *Voyage dans la Basse et la Haute Égypte*, 21
el Berberi, Hassan / Khassan, 38, 40, 45, 49–50, 52, 55, 62, 75, 80, 112, 122–3, 134, 136
Emesa. *See* Heliodorus
England, 17–18, 25, 31, 90, 92, 113, 126, 134, 137, 145, 159n5, 161n4; citizens, 31; consul (*see* Salt, Henry); court, 31; Expedition of 1807, 21;

INDEX

forces, 125; giraffe, *18*, *35*, 126, *127*, *128*, 129, 145; giraffomania, 126; king of (*see* George IV); oaks, 126; technician, 126

Epidaurus, 25

Épinal, 93

Eritrea, ix

Essonne, 93

Estienne du Bourquet (d'), Louis-Jules (Mayor), 77

Étampes, 15-6, 20

Ethiopia, x, 10, 34, 121; deserts, 55; giraffe, 51; people of, 1

Europe, 10, 21, 31, 33, 42, 49, 55, 63, 92, 97, 112-113, 115, 125, 132, 144, 151; naturalists of the Middle Ages, 3; people, 37, 149, 159n2; Western, 27; zoos, 147

Europe 1, 153

Ferrand, Humbert, 28

Field Museum of Natural History, x

Flandrin, Auguste, 100

Florence, 10, 49

Fontainebleau, 85-86, 93, 97

Fontenay fields, 124

Fourier, Charles, 148

Francis I (king), 63

Frayssinous, Denis (count, bishop), 69

Frederick II (king), 3, 10

Gaspard, 114

Gay, Jean-Baptiste (viscount of Martignac), 137

Géménos, 51, 69-70

Genoa (Gulf of), 63

Geoffroy Saint-Hilaire, Étienne, 13, 15-20, 60, 66-70, *67*, 72-81, 83-93, 97, 112, 116, 118, 121, 125-6, 129, 133-4, 136, 144, 146-7, 149, 152-3

Geoffroy Saint-Hilaire, Isidore, 69, 91-92, 146-7

Geoffroy Saint-Hilaire, Marc-Antoine, 16

George IV, 31-35, 126-9, *127*, *128*, *129*; mistress (*see* Conyngham, Elizabeth)

Germanos III of Old Patras (bishop), 25

Germanic emperor. *See* Frederick II

Germany, 25, 47

Gesner, Conrad, 4

Gibraltar, 63

Gien earthenware, 101

Giraffa, 3

giraffe: Abyssinian, 3, 147; *camelopardalis*, 2-3, 68; Ethiopian name, 1; Greek name, 2-3; *hippardion*, 2; Italian, 10, 145; Kordofan, x; Latin name, 3-4, 154; *Maasai*, 154; Nubian, x, 154; Orafle, *oraflus*, 3-4; *ovis fera*, 3; *pardion*, 2; *reticulated*, 154; Sennari, 74; South-African, 154; *Zemer*, 1

Gondelier, 113

Gordian III, 3

Goritz, 145

Greece: conflict with Ottoman Empire, 15, 25-29, *26*, 48, 109, 112, 114-116, 125, 134, 138, 144; *Greece on the Ruins of Missolonghi* (*see* Delacroix, Eugène); people of, 2, 7.

See also giraffe

Guib-Allah (male giraffe), 146

Guinea, ix

Gujarat (king of), 63

Hamilton, William Richard, 17

Hardy, 119

Hassan. *See* el Berberi, Hassan

Haüy, René Just (abbot), 16

Heath, William, 126-8, *127*, *128*, *129*

Heliodorus of Emesa, 2

Hermonthis, *30*

hieroglyph, 2, 148

Hohenstaufen dynasty, 3

Hollywood, xii

INDEX

Horace, 3
Hottentots, 59, 103
Hugo, Victor (*Les Orientales*), 29
Hussein (dey), 97, 113, 137, 143, 148

Ibn al-Faqih, 3
Ibrahim (Pasha), 22, 27, 137
I Due Fratelli. *See* Manara, Stephano
If (island of), 63
Île-de-France (museum), 153
India, ix; cows, 136; elephants, 137, 151
Indians. *See* Osages
In Djaren wadi, 1
Indochina, 152
Institut Pasteur, 153
Isère, 79, 81
Istanbul, 106; Sultan of (*see* Mahmud)
Italy, 145; campaign, 20; French Consul, 13; librettist, 27; minister of Tuscany, 93

Jacobinism, 31
Jacquemin, 133
Japan, ix
Jardin des Plantes, 31, 48, 97, 102
Jauffret, Adolphe, 103
Jazet, Jean Pierre Marie, 32
Joigny, 91–92
Jomard, 30
Jonston, John (or *Jan*), 7, 8
Joly, Nicolas, 5, 30,
Julius. *See* Caesar
Jupiter, 2
Jussieu cedars, 99

Kenya, x; coasts, 151
Kepler, Johannes, 7
Khartoum, 15, 34, 60
Kihégasuhgah, Lord (fictional character), 130
Kilian, Conrad, 1
kilin, 4

King's Garden (zoo), 38–39, 49, 70, 72, 74, 78, 81, 92, 97, 99, 102, 106, 109, 113, 115, 118, 121, 123–5, 130, 132–133, 135, 137, 140, 144, 146–148
Kléber, Jean-Baptiste (general), 17
Kordofan, 15, 51; desert, xi
Kurd(ish). *See* Ali, Muhammad

Laborde, Joseph de, 137
Lacépède, Bernard-Germain de (count), 16
Lacroix, Alfred (professor), 152, 154
Lacuée, Jean-Girard (general), 43
Lady Conyngham. *See* Conyngham, Elizabeth
Lafaille (museum), 151
La Fontaine, Jean de, 104
Lamarck, Jean-Baptiste de, 68–69, 149–150
Lambesc, 77, 79, 81, 85
Lancaster, 126
Langlumé, Pierre, 133, 138
La Palud, 71, 84–85
La Provence (ship), 138
La Rochelle. *See* *Muséum d'Histoire naturelle de La Rochelle*
Latin. *See* Giraffe
La Truite (barge), 39, 42
Laveau (count of), 21
lazaret, 37, 41–45, 47–48
Lebanon, 152–153
Lebon naural gas, 101
Lebrun Gallery, 27
Legion of Honor, 22, 135
Le Havre, 63–64, 129
Le Huen, Nicole (Brother), 4, *5*
Leo Africanus, 4
Leonidas, 29, 115
Leo X (pope), 63
Les Deux Frères (ship). *See* Manara, Stephano
Les Invalides, 147

INDEX

Les Islettes earthenware, 101
Les Orientales. See Hugo, Victor
Levaillant, François, 10, *12*, 48, 58, 147–8
Levant. *See* Eastern Mediterranean
Lhote, Henri, 1
Liberia, 151
Libyan desert, 153
Lie, Jean-Christophe, xii
Limoges porcelain, 101
Lipparini, Ludovico, 26
Lisbon, 18
Livourne, 38
London, 25, 35, 126, 145; Treaty of, 113; zoo, 145–146; Zoological Society of, 129
Loppé, Étienne (doctor), 151–2, 154
Lord Byron. *See* Byron, George Gordon
Loriol, 71, 84–85
Lot-et-Garonne, 43, 92
Louis-Philippe I (king), 144–5, *145*, 147
Louis XI (king), 10
Louis XIV (king), 89–90
Louis XVIII (king), 20–21, 31
Louvre, 144; museum, 33, 134
Luxor obelisk, 137, 144–5
Lyautey, Hubert (marshall), 152
Lyon, 4, 71, 84–86, 88–91, 99, 114, 144; École nationale vétérinaire de, *6*, *8*
Lyons, 71

Mâcon, 90
Madagascar, 152
Madame the Dauphine See Angoulême (Duchess of)
Madame Royale. See above
Mademoiselle. *See* Berry (Duchess of)
Magnus, Albertus, 4
Mahm(o)ud (sultan), 26–28, 38, 113, 134
Maisse, 93
Mali, ix
Malta, 126

Mamluk(s), 10, 13–15, 73. *See also* Qaitbey
Manara, Stephano / Stefano (captain), 38, 39
Manfred (king), 10
Manuel the Fortunate (king), 63
Marie-Antoinette (queen), 90
Marignan, 63
Marmol, Luis del, 7
Marseille, xi, 20, 31, 37–42, 44–5, 48–51, 53, 55, 63–4, 66, 70–81, 83, 85, 89, 91, 97, 103, 105–106, 112, 122, 134, 136, 159n4, 159n7, 159n1; Académie de, 83; *Collège royal de*, 50
Martignac. *See* Gay, Jean-Baptiste
Martin (bear), 115, 140
Matherou, Philippe, 53
Mauritania, ix
Mecca, 15, 116, 118
Medici, Lorenzo de', 10, 49
Medina, 15
Mediterranean (sea), 15, 39, 146; Eastern, 39; French coastline, 42
Melun, 93
Ménagerie. See King's Garden
Méry, 48, 50
Metternich (von), Klemens, 113
Meuse (river), 151
Michelet, Jules, 19
Middle Ages, 3
Middle East, xi, 153
Minerva's temple, 115
Ming dynasty, 4
Minister of the Colonies. *See* Reynaud, Paul
Minister of Foreign Affairs. *See* Damas, Ange Hyacinthe Maxence de
Minister of the Interior. *See* Corbière, Jacques-Joseph (count)
Minister of the King's House. *See* Villèle, Joseph de
Minister of War. *See* Bourmont, Louis de

INDEX

Mississippi, 133
Missolonghi, *26*–27, 115
Missouri, 133
Mohammed III, 7
Moloch, 29
Mongez / Mongès, Antoine, 97, 123
Monsieur the Dauphin. *See* Angoulême, Duke of
Montastruc mill, 92
Montélimar, 71, 84–85
Montereau, 85–86; earthenware, 101
Montesquieu, Charles Louis de Secondat de (baron),123
Mont Redon, 51
Morea, 27, 137
Moscow (Imperial Society of Naturalists), 21
Mouker Bey of Sudan (governor), 31, 34
Muhammad. *See* Mahmud
Murat, Joachim I (king), 21
Muséum central des Arts, 22
Museum of Egyptologie. *See* Turin
Muséum d'Histoire naturelle (Fleuriau) de La Rochelle, xi, 11, 12, 151–154
Muséum national (royal) d'Histoire naturelle (Paris), xi, 16–20, 22, 33, 44–45, 47, 49, 52–53, 61, 64, 66, 69–70, 72–73, 77, 19, 81, 83, 85, 91–92, 99, 125, 129, 132, 144, 146–147, 149–4; *Orangerie*, 112

nabis / *nabu* / *nabuna*, 1
Naples, 43
Natural History Faculty. *See Muséum national d'Histoire naturelle*
Navarin (battle of), 125
Navarre. *See* College of Navarre
Near East, 154
Nevers earthenware, 101
Nicolas I (czar), 113
Nigeria, ix

Nile (river, delta, valley), 15, 22, 34, 58, 60, 121, 146
Noah's Ark, 38, 151
Normandy, 19
Nubia, 53; sheep, 21

Odessa, 25
Odevaere, Joseph-Denis, 26
Oglou (countess of), 119
Oppian, 2
Orange (river), 48
Orange (town), 71, 83, 85
Orgon, 79, 80, 85
Oriental Expeditionary Force, 17, 20
Orléans monarchy, 144
Osages, 129–30, *130*, 132–3, *133*
ossicones, ix, 48
Ostende, 133
Otto I (king), 144
Ottoman. *See* Turkey

Paganini, Niccolò, 19
Palazzo Vecchio, 10
Palermo, 4
Palestine, 134
Pamphili, Eusebius, 3
Panckoucke, Charles Louis Fleury, *30*
Paré, Ambroise, *6*
Pasha. *See* Ali, Muhammad
Pellerin, Jean-Charles, 93
the *Penelope* (ship), 126
Péri-Courcelle, 92
Perrault, Charles, *46*
Perrier, Casimir, 61
Petitbon, 119
Peyronnet, Pierre-Denis de (count), 61
Philhellenism, 31, 112; disciples, 25, 29, 81, 115, 144
Philip the Arab (emperor), 3
Philosophie zoologique. See Lamarck
Piedmont, 20, 22
Place de la Concorde, 145

INDEX

Pliny the Elder, 3
Polastron, Louise de, 31
Polignac, Jules de, 138, 143
Polito, 64–66
Polo, Marco, 3
Pomègues (port), 39
Poortman, 150
Porte d'Aix, 44
Port Said, 153
Portugal, 18, 85; king of (*see* Manuel the Fortunate)
Provence, 88; people of, 81. *See also La Provence*
Pyrenees, 10

Qaitbey (Mamluk), 10
Quai d'Orsay, 152
Quélen, Hyacinthe-Louis de (archbishop), 144

Redon. *See* Mont Redon
Redouté, Henri-Joseph, 17
Red Sea, 15
Reims, 33
Renaissance, 3–4
Restoration, 20, 152
Revolution: of 1789, 33, 135, 149, 158n2; of 1830, 143–4. *See also* Greece
Reynaud, Paul (minister), 152
Rhône: river, 64; department, 90
Richelet, César-Pierre, 7
Rif'â-at-Tahtawi, 42
Robert (fictional name), 114
Robida, Albert, *46*
Rochefoucauld, Sosthène de la, 28
Romanzoff (count), 21
Roman roads, 90
Romantic, 104; painting, 26; poet, 29
Rome, 3, 63, 83, 103
Rossini, Gioachino, 27
the Rotunda, 77, 135

Royal Zoo. *See* King's Garden
Russia, 25, 92, 113, 134; army, 125, 137; chancellor of, 21; lord, 125

Sahara, 1, 51, 78
Sahel (Central), x
Saifuddin Hamza Shah (king), 4
Saint-Andiol, 79
Saint-Cannat, 79, 81
Saint-Cloud (royal castle), 116–7, *117*, 121–3, 133; *La Girafe à*, 119; *Lettre de la Girafe…*, 123
Sainte-Pélagie prison, 99
Saint-Firmin prison, 16
Saint-Lambert, 85
Saint Louis (king), 3
Saint-Symphorien, 84, 85
Salt, Henry, 31, 33–34, 126
Salvandy, Narcisse-Achille de (count), 123
Salze, 55
Sand, George, 19
Saône (river), 84, 90
Saône-et-Loire, 90
Sardinia: ship, 38–39. *See also* Charles Felix
Savigny, 17
Savoy, Duke of. *See* Charles Felix of Sardinia
Sayyidah Zaynab, 34
Sceaux (museum, castle), xi, 153
Schönbrunn Zoo, 137
Seine: river, 63, 92–93; -et-Marne (department), 92
Senegal, ix
Sennar, 34, 53, 58, 60, 97, 112, 123; governor of (*see* Mouker Bey)
Sève, Joseph Anthelme, 26, 134
Seymour, Robert, 126
Sicily, 10
Sidi Ferruch, 143
Singer, 102

INDEX

Soliman Pasha. *See* Sève, Joseph Anthelme
Solomon, 143
Somalia, x
Sorbonne (university), 19, 146
Soult, Jean-de-Dieu (marshall), 144
Soumet, Alexandre, 27
Spain, 18, 41, 85; writer, 7; castle, 51
Sparta, 29
Statistique du département des Bouches-du-Rhône. *See* Villeneuve Bargemon, Christophe de (comte)
Stendhal, 93, 144
St. Helena, 147
Sudan, 15, 21; East, 154; governor of (*see* Mouker Bey); *Marble*, 102; South, x. *See also* Atir
Suez: Canal, 151, 153; desert, 152; port, 15
Sumatra, 151
Syria, 134, 152

Tain (l'Hermitage), 84–85
Tarasque, 45
Tassili, 1
Taylor, Isidore (baron), 145
teleausorus, 19
teratology, 20, 69, 147
Theagenes. *See* Heliodorus
Théaulon, 113
Thévenard, Pierre (professor), 153
Thiers, Adolphe, 144
Tonnerre, 93
Toulon, 72, 75, 136, 143
Toulouse, 147
Tripolitsa, 25
Troglodytes, 2
Turin, 20, 150; Museum of Egyptology, 22
Turks / Turkeys, 112, 114–5; army, 137, 143; conflict with Greece, 15, 25–7, 29, 38, 48, 81, 109, 115, 125, 134, 137–138; empire, 15, 92; pashas, 13; sultan (*see* Mahmud); wheat, 71
Tuscany, minister of. *See* Berlinghieri, Deniello

United States, 25

Valence, 71, 84–85
Valley of the Kings, 34
Vasari, Giorgio, 10
Vaucluse, 79, 81
Vaudeville, 113–114
Venice, 137; Film Festival, 153
Verdun: museum, 151–152, 154; trenches, 151
Verreaux, Jules, 151
Versailles palace, 135
Vial (doctor), 63
Vienna, 21, 137; porcelain, 137
Vienne, 71, 85–86
Villefranche, 90
Villèle, Joseph de (minister), 33, 37, 49, 64, 75, 81, 83, 85, 90, 109, 111, 113, 118, 121, 123–124, 137
Villeneuve Bargemon, Christophe de (comte), 39–45, *43*, 47–52, 61–66, 69–70, 72–74, 77–81, 83–85, 90–93, 99, 121–123, 129, 133–134, 136–137, 140, 153; *Statistique du département des Bouches-du-Rhône*, 39, 44, 74, 83; Wife (countess), 74, 78, 123
Villeneuve-Saint-Georges, 93, 97, 112
Vincennes. *See* Bois de Vincennes
Vitry, 92
Volo (Gulf of), 144
Vosmaer, Arnout, *8*

Wahhabites, 15
Waterloo (battle of), 19
William IV (king), 129

INDEX

Windsor, 35, 126
World War I, 151
World War II, xi
Würtemberg (king of), 21

Yonne, 93
Youssef. *See* Ebed, Joseph

Zaida (female giraffe), 146
Zanzibar, 3
Zaynab. *See* Sayyidah Zaynab
Zemer. See giraffe
Zoo garden. *See* King's Garden
Zoological Society. *See* London
Zhu, Di, 4

ABOUT THE AUTHOR AND TRANSLATOR

ABOUT THE AUTHOR

After his English and media studies, **Olivier Lebleu** worked for independent radio, written press, and then television production. The award-winning author of several historical narratives and novels, he has also written a musical comedy, plays for theater, dramatic performances, songs, and film scripts. In addition to writing his own material, he is an English-to-French nonfiction translator.

In 2002, when Lebleu moved to La Rochelle, France, he "encountered" Zarafa in this city's Museum of Natural History, where the stuffed skin of this nationally recognized animal has stood since 1931. His 2006 book, *Les Avatars de Zarafa*, rekindled some giraffomania in France. In 2012 the children's animated film *Zarafa* drew from his research and opened the subject to a wider audience. Since 2016 Lebleu has regularly played a role in his dramatic piece, *Le Talisman de la Girafe*. He has been giving lectures on Zarafa's story across the world for more than a decade.

ABOUT THE TRANSLATOR

Cynthia T. Hahn, PhD, has been teaching French and Francophone language, literatures, and cultures as well as creative writing and translation at Lake Forest College, Illinois, since 1990. Her book-length creative publications include ten translations of novels, volumes of short stories, and poetry by French, Algerian, and Lebanese writers, as well as two original volumes of poetry in English and French.

www.ingramcontent.com/pod-product-compliance
Lightning Source LLC
Chambersburg PA
CBHW070330230426
43663CB00011B/2268